Backbone

by

George F. Brannon

Brannon, George F.
Backbone / George F. Brannon.
p. cm.
Preassigned LCCN: 95-94430
ISBN 0-9645309-0-2

1. Brannon, George F. 2. Vietnamese Conflict 1961-1975--Personal narratives, American. 3. Vietnamese Conflict, 1961-1975--Veterans, United States--Biography. 4. Loggers--United States--Biography. I. Title.

DS559.5.B73 1995 959.704'38
QBI95-20245

Published by Brannon Enterprises, Inc.
527 North Opal Street
Sutherlin, OR 97479

Production by Frontier Publishing
Seaside, OR 97138

Printed in the United States of America

Preface

Today our nation faces mental—as well as physical—destruction of families and individuals. Much of the fault is placed on what are perceived as diminished family values, child abuse, neglect, ineffective education systems, lack of respect for law and government, illegal drugs, negative effects of welfare, and other adverse factors of today' living.

Backbone is the true story of a man, George "Bud" Brannon, who was born into the period considered the genesis of these modern-day problems. Small of physical stature, deserted by his family, thrown into the roughest of occupations, drawn into the nation's most unpopular war, he applied brawling and hard drinking as weapons to conquer a life he had not asked for.

He gained little until he realized these experiences were revealing that God was most important. In time he found that his sole strength was faith in God through His son, Jesus Christ. The revelation brought him peace.

Dedication

The author dedicates *Backbone* to all the lost young children searching and trying to find their way, the working man, and the Vietnam Veterans who willingly gave all they had to represent their country.

Some lost their lives, some were injured and some came home. But they all are truly the backbone of our country.

May they find peace and strength to continue God's will.

About the Author

George Brannon was born in Pryor, Oklahoma, in 1943, but has spent most of his life in Oregon.

After graduating from high school in a small Oregon town, he got his first job in the woods. He served in the Armed Forces and spent time in the Vietnam War. He was wounded during the war and received a Purple Heart and Bronze Star.

Following his honorable discharge the author returned to work in the logging industry.

He has cut timber in Oregon, California, Washington, Idaho, Nevada and spent several years in Alaska.

Table of Contents

Chapter 1
Going West 1
Chapter 2
Beginnings of a Scrapper 7
Chapter 3
On to Oregon 21
Chapter 4
The Family Fades 28
Chapter 5
Logging Alaska Style 37
Chapter 6
Northern Adventure 41
Chapter 7
Mexico: Foes and Friends 65
Chapter 8
Fights, Cops and Frisco 73
Chapter 9
Vietnam Beckons 89
Chapter 10
Learning War 105
Chapter 11
The War Gets Real 118
Chapter 12
An Attack That Worked 130
Chapter 13
Pinned Down 136
Chapter 14
One Battle Too Many 143
Chapter 15
Hospital, Medals and Out 148

1

Going West

On June 5, 1943, in Mayes County, Pryor, Oklahoma, the sky was getting dark with graying clouds building up in the northern part of the state. Then suddenly, the wind hit, with driving rain coming just minutes behind.

It came fast and hard. It sounded like a tornado had almost hit, the way the rain and wind whistled through the cottonwood trees on the surrounding hills and valleys.

Just a little ways off from Pryor, in the rolling foothills, stood a three-room shack. Inside was a bed, rocker, table, three wooden chairs and an old wood-burning cookstove. Also inside were two people, a pretty young woman in the last month of the family way and her young daughter.

They had no sooner gotten dressed when the door flew open. It was her husband helped by the wind that also drove sheets of rain in behind him. He had an armload of wood he had just gathered up off the front porch. Since he was soaked to the skin from the sudden rain, he said a quick good morning, dropped the wood by the stove and headed for the bedroom to change his wet, dripping clothes.

In just a few minutes, he came out again and started putting paper and kindling in the stove. Before long, he had a red hot fire going. Since the cabin was small, it didn't take long for it to warm up.

After sitting down for a cup of coffee and breakfast, he was about to go to bed, when his wife started having labor pains.

Lucky for him, the neighbor lady had gotten up early and had decided to come by and check on his wife. JoAnn was a good friend of hers and also a mid-wife. George Dewey Brannon then headed back out in the sheets of rain driven by the wind, to get a doctor.

In the last three to four hours, the rain had come down so hard that the road crossing to Pryor was covered with water deep enough it was impossible to cross. He then headed southeast for about a quarter of a mile where he was able to wade the swollen creek.

George then headed north to a farmhouse where he was able to use their phone to call a doctor. He waited for old Doc Davis and they headed back to the house.

Doctor Davis was driving his car, so they went back on the road which was longer but had a bridge across Pryor Creek. Three to four hours later, they finally arrived back at the house.

Upon entering, George and old Doc Davis found Una Juanita Brannon and JoAnn sitting on the bed playing with little, newborn George Floyd Brannon (the author). The doctor did a quick check just to make sure everything was all right. He noticed a piece of regular string had to be used to tie the navel cord.

JoAnn told the doctor that she had to use a string holding a curtain on the window. Old Doc started laughing, and soon George was inquiring to find out what was so funny. Doc Davis said to him, "I can see these women had it all under control."

As the spring went by, the hills and valleys seem to fill out with color and almost came alive. Mrs. Una Juanita Brannon was now up and around and fully recovered from giving birth to her new son. In fact, the young baby boy was not only a full-time job for her, but also made a good companion for her daughter, Stella Lois Brannon.

Stella loved playing with him so much, she called him her little "Buddy." From that time on, everyone started calling George Floyd Brannon, "Buddy."

As summer went on, the little boy continued to grow, and then autumn came with its many colors of leaves. The days started getting shorter and winter came.

Along with winter came the cold, rain, and winds. Before the family realized it, it was spring again and then summer. Time seemed to fly, it was going by so fast.

The little boy had colic and almost died the last fall. Finally, the young woman found goat's milk and that was the only thing he seemed to be able to hold down. He was now three years old and it was almost spring again.

Times were hard because the powder plant had shut down. World War II had ended and there just wasn't enough work for a man without a trade or some kind of degree. The young man and his wife decided to go out to his brother's place in Turlock, California, when spring came. His brother had written and said that there were lots of jobs if a man was willing to work.

Spring came, with the hills turning green and wild flowers of all kinds spotted all over the landscape. It made a pretty, picturesque scene.

The day had finally come to make the move. My mother was standing on the front porch looking out over the valley with tears in her eyes. She was talking to the old mid-wife, JoAnn, and relatives. I remember how sad she was because of moving away from all the relatives and friends she had known since she was a little girl.

Being a young boy, I couldn't understand why we didn't get going sooner. It was exciting to me, getting ready to go on a long trip. After all the relatives shook hands and talked with my pop, he said it was time to go. Dad and Mom put the last of the luggage in the trunk of the old 1937 Chevy sedan and we were off.

The first day was really exciting; the four of us traveled across the rolling hills of Oklahoma. My sister and I watched out the windows, full of excitement, as we headed for the new home in Turlock. We watched the dust of the old road behind the Chevy as we raced across the old country roads.

Towards evening time the excitement started to wear off, and fatigue, caused by the long day of riding, set in. Soon night fell and my pop started tiring and decided to find a place to stay. After finding an old adobe motel, Dad went in to get a room to spend the night in. Then he came back out with the key and unloaded our luggage with Mom's help. They came back for me and my sister. Dad carried my sister and Mom carried me, because I was smaller. The room was not very big, but clean, with an old toilet that flushed by pulling a string.

The morning came soon as my dad wanted to get an early start. My sister and I were told to get up and make our rounds to the bathroom so we wouldn't be wanting to stop all the time. After that, Dad and Mom loaded our luggage and we all loaded up for another long day of traveling. Both of us kids were feeling fresh and rested; we were ready for the exciting trip.

We were so fascinated by the beauty of the scenery along the road side. We would just sit for a couple of hours at a time staring out the windows. Then, like normal kids, we started pestering each other and before long we were fighting.

Dad sternly told us to knock it off or he would be stopping the car. Both of us knew he clearly meant business and we would stop talking to each other for awhile. That satisfied Mom and Dad; they knew they would be able to enjoy some peace and quiet. After about an hour, we were playing once again as if we had completely forgotten about our petty argument.

It was time to stop and get lunch. My sister and I really enjoyed this part of the trip. We had never seen a restaurant and were excited about being able to go in and order what we wanted to eat. The first day, my mother had made sandwiches and something to drink, so we wouldn't have to stop.

Dad found a nice restaurant and soon we were inside being seated and given menus. We had never had a coke or eaten anywhere other than with friends or relatives. The waitress was the owner's wife and she could quickly see how my sister and I were taken up with all that was going on. She made sure we got special attention and even gave us a little candy treat when we were done eating.

After paying for our lunch and saying thanks, we were once again loading in the car to continue our long trip. Soon my sister and I were settled in and playing again.

We weren't on the road long before we were into Grand Canyon National Park. We were fascinated looking at all the rock formations and different colors in the rocks. Soon evening came and my sister and I were starting to doze off to sleep.

It only seemed like a few minutes and I heard Dad tell Mom he needed to stop and get some rest. But instead of going to a motel he just pulled off the side of the road to sleep, since we were in the middle of the desert. Mom covered my sister and I with blankets and settled down in the front seat to catch a little rest herself.

After a few hours of sleep, I was quickly awakened by the movement of the old sedan pulling back onto the road. After a few minutes, my sister started to slowly awaken. All of a sudden, we were driving up and down a lot of hills and Dad said we had finally reached California. Lois and I were really excited because, by this time, we were getting tired of sitting and riding for hours and hours.

We didn't realize we still had several hours' ride ahead of us. Before it was over Dad wished he hadn't told us that. My sister and I kept asking him if we were almost there. Finally he said that he would let us know when we got close. It seemed like a very long time, but finally he said we were only a little ways off.

2

Beginnings of a Scrapper

Soon we were driving into a small town with tall palm trees everywhere. My sister and I could not see what was so great about this place, but before long we thought it was the greatest. This was because Dad's brother and his wife had a boy and a girl almost the exact same age as us. Their girl was almost the same age as my sister and the boy was almost the same age as me.

Dad found work and then located one or two acres out of town he could buy pretty reasonable. After closing the deal he started getting the ground ready to build on. I remember living in a big Army type tent while Dad dug a septic tank hole, a well, and started building a house.

He dug the well first because we needed water right away for drinking and cleaning up. First, he took a post hole digger and dug till he got too deep for the pipe he used to extend the digger handle. Then he pulled up the post hole digger and hauled water to fill the hole. Next he attached another piece of pipe to make the digger handle longer and drilled until he ran out of pipe again. When he got to a certain depth, he put the casing in to keep the post hole digger going in a straight line.

I remember him hitting water. He sure was happy

about that because, being summer, it was hot, hard work and he had to do it at night and on his days off from his regular job. Mother was happy; she didn't have to pack and ration the water for washing clothes and bathing us.

Next, Dad started the septic hole. He finally got it almost finished when the boy next door and I got into a fight. The boy was about two years older than me, and he was bigger. He whipped me pretty good.

Several minutes later Dad came home. Since the boy had already gone home, I went crying to Dad. This I found was the wrong thing to do. Instead of consoling me he sat me on his lap and said he was going to spank me. Then I really started crying, but he said, "It's just because I want you to learn to handle things by yourself and not to look for someone's sympathy." He told me he might not be around someday so I had to learn to stand up for myself.

At that Dad took his belt off and bent me over his knee and gave me two or three swats on the behind. He also gave me a few pointers on what to do in a fight, because the kid was bigger than me. I was sobbing but he then sat me up and told me I was going to have to go get the boy so he could beat me up again. Even though I didn't want to, after getting his advice I headed over the field to the boy's house. I remember walking slowly through the wild grass.

As I walked up the steps to knock it seemed like I had gotten there too quickly. Soon the door came open and the boy, Roger, asked what I wanted. I told him my dad wanted to talk to him.

He asked, "What about?"

I said, "I think he wants you to whip me again."

Roger gave me a puzzled look and then said okay. He

yelled to his mom that he had to run next door for just a little bit.

She yelled back, "Okay, but don't be too long." Then he and I headed for my place.

When we arrived Dad was sitting on the running board of his old truck he had traded our car for. Roger then asked what he wanted and Dad told him to whip me again.

He told Dad he really didn't want to, but Dad said to do it anyway. Roger then said okay and then he turned around and started for me.

The battle was on but this time it was pretty even up. Finally Pop asked us if we both had enough; if not to keep fighting.

We both kept on swinging and rolling around in the dirt for several more minutes, putting on some bumps and bruises, and blackening each other's eyes.

Then Dad asked us again if we wanted more. By this time both of us had had enough fighting. He made us shake hands and told us if we ever wanted to fight again to wait till he was there to make sure we got it out of our systems.

After that Roger went home and I felt much better for standing up to him even though he was bigger. Dad told me anyone could be whipped if you went about it right. He said speed and leverage will whip muscle and that's why he figured I got the best of Roger. I didn't feel like I got the best of him but if Pop said I did, it was good enough for me. With this lesson, I never again went looking for anyone's sympathy after a fight. Especially my dad's.

Roger and I became real good friends from then on. He was over at our place all the time. When my dad got home from work, we would all go swimming and Roger would almost always go along.

That first summer in California was a real summer to remember. Then Roger moved, and school started when fall came. This was a lonely winter, with my sister starting school and Roger moving. I was left to myself most of the time.

So one day Dad came home with a big surprise. He was over at the house of a friend who offered to give him a dog for me and my sister. The dog was part coyote and part German Shepherd. It looked more like a coyote than anything. I asked Dad what I should name him. After a bit, he said, call him Lobo.

From that day forth Lobo went everywhere I went. When my sister was home, he went with both of us. Dad trained Lobo to ride on top of the truck when we went to town. He would jump on the running board, then the fender, hood and onto the top. He stood on top like he was glued to it.

Lobo only fell off once and that was because Pop had too much to drink and took a corner too fast. My sister and I were crying and screaming our heads off because he was driving so fast and was going to kill Lobo or hurt him bad. It seemed Pop had been drinking a lot, so my sister and I started having him leave Lobo home. Lobo was good at watching the place, so Dad really didn't mind leaving him.

Soon hunting season came and Dad decided to go hunting. He bought a bigger truck that was covered on the back for a place to sleep while we were out in the mountains.

There were big trees all over and all kinds of animals. We even saw a herd of about twenty-five deer. We stopped to look at the deer and Lobo jumped out and started chasing them. Dad was drinking again and got mad at Lobo. We tried to call him back, but couldn't get him to come back, so Dad said to hell with him and drove off. My sister and I started crying again, but soon we shut up. Mom would tell us that when Dad was drunk, we had better be quiet before he got mad.

Mom tried everything she could to keep us from worrying about Lobo. She even told us he would find us or come home when we got home. I knew she was only trying to make us feel better.

After the second day, it was time to go home, and Lobo had still not come back. My sister and I just knew we would never see Lobo again. Not only that, we never even got a deer. I remember I didn't care if I ever went on a hunting trip again. Not only that, I hated going anyplace with Dad when he was drinking; he did scary things and he scared Mom and my sister.

After getting home I was awfully unhappy and lonesome without Lobo when Sis went back to school.

About six to seven weeks went by and, on a sunny day when I was out playing, I looked up and Lobo was coming across the field. I started jumping up and down and yelling. I must have scared Mother. She came running out to see what all the noise was about.

When she got outside she saw me running across the field to Lobo. Then she noticed Lobo looked like he was hungry and his feet were sore, because he was limping and looked awful skinny.

Mom came running also. When she got there I was hugging Lobo around the neck and kissing him. She dropped down on her knees and started hugging me and Lobo. Then after a bit she said we better take Lobo over to the house and feed him.

She fixed him some warm milk. Mom said we better not try and give him too much at first. After that she doctored his feet by putting them in some kind of warm solution. When she was done she told me not to bother him too much; he could use the rest.

I couldn't hardly stand it, so I just sat and stared at him for hours. Pretty soon Mom yelled for me to come in and eat something. I really wasn't hungry so I yelled that I wasn't. I really didn't think she was going to let me go without eating.

Several minutes went past and sure enough, here came Mom with a sandwich. She told me I had to go wash my hands and face before I could eat. I was moving a little slow because I really didn't want to leave Lobo, but she said hurry and I'll watch Lobo for the time it takes you to wash your hands and face.

I was still feeling hesitant, but off I went to the new well Dad had dug. After pumping on the handle a couple of times, the water started to come out. I quickly splashed my hands and face and ran over to mom for an inspection. She looked a little bit and said it wasn't very good, but would pass just barely. Then she handed me my sandwich and went back inside.

After she had been inside a while, I decided Lobo must still be hungry and I wasn't a very big eater, so I gave most of the sandwich to Lobo. He must of been hungry, it was gone in a couple of gulps. He started licking my hands, just as if he was trying to thank me. It was so great to have him home again. "Wait till I tell my sister," I thought.

Before long I could see the kids coming home. Mom came outside and took one look at me and said go ahead. Let me tell you, I took off down that road as fast as my legs could carry me. I couldn't wait to break the news to my sister. I always ran to meet her, but this time, it was special. When I got just about to her, I yelled, "Lobo is home! Lobo is home!"

She was so excited, she started running as fast as she could for the house. She could always out-run me, but today she really left me behind. When I got there she was hugging Lobo. Lobo was equally happy to see her. His tail was going pat, pat, pat on the old hard ground. After a minute or two, Mom came out and told Lois she had better go in and change her school clothes so she wouldn't ruin them. After changing her clothes, she quickly returned to love Lobo.

Mom had told her not to get him too excited so he could rest, so my sister and I just sat there looking at him and then we started to talk about how Lobo had come so far, and how did he find his way home?

Soon Dad came home and when he saw us with Lobo, he could hardly believe his own eyes. Dad said that Lobo must really love us kids to come through all those mountains and towns so far away. We asked him just exactly how far it was and how he could've possibly made it back. He said it had to have been two to three hundred miles and that some dogs just have a good instinct for finding their homes.

That night everyone was in such a good mood and to make things even better, Pop hadn't been drinking.

The next day when I got up, Lobo must of been feeling better because he started his usual trip to school with my sister. After a little bit he would come home again. Then, when it was about time for her to get out, he would go back to school to wait for her. As soon as class let out he would be lying outside the door waiting to walk her home.

It was about a mile to school. Mom and Dad told her to come straight home. I always knew when Lobo left that Sis would come home soon. After Lobo went to get Sis, I would run and ask Mom if I could go meet her. Mom always said yes, but to be careful and to stay off the road if a car came along. I would run down the driveway and turn left on the old dirt road. I would run as far and as fast as I could, until I was panting so hard I had to slow down to almost a walk, then I would take off again.

I did this all the way to a little store on the route. When I reached the store, I would see Sis and Lobo just a short ways off. They would come running to see if Mom had given me any pennies to buy some one-cent candies in the candy jar. Lobo also looked forward to the sweet treats. Sis and I would always break off little pieces and give some to Lobo.

Sometimes we even got a nickel each to buy a Coke and sometimes one of us would buy a Coke, and the other would buy some candy. That way Lobo got something, too. Then we would walk home, taking turns drinking out of the Coke bottle and munching on candy. We would check in at the house before going out to play.

After playing a while, we would have to go in and get cleaned up for supper, and for the next day. Sis and I liked playing in the tub of warm water, but I hated Mother washing my hair, face and ears. It seemed like I always ended up crying because she would get soap in my eyes and scrub me too hard. My sister always seemed to get to do her own scrubbing. I would tell Mom I would sure be glad when I could wash myself. She would tell me I could now if only I would do it instead of playing. Soon Dad would drive up and it would be time to get out of the tub and get dressed.

When we were dressed, Mom would set the table and we would all sit down and eat. After eating, Mom would help Sis do her homework. Then we would have a little time to play before going to bed. After getting ready for bed we would always play on the bed until Pop would say that's enough and tell us we had better get to sleep because Sister had to get up early for school. By this time I was usually worn out from the day's activities and would go fast asleep in just a few minutes.

Soon spring came. The orchards bloomed and the fields turned green and were full of birds eating and chirping. I really liked this time of year because I knew summer wasn't far off. Then I would have someone to play with and we would be able to go swimming when Dad got home from work.

Summer came almost sooner than I had expected. That summer was really lots of fun for me and Sis because Lobo got to go along. Dad trained Lobo to jump in the water and pull us back to the canal bank by hanging on to his tail. This was fun except when Dad would get drunk and throw me in.

I remember one time he threw me in too close to a partly open canal gate. Before Lobo could get to me the water sucked me down under and into the tube the water was channeled into. Dad's friend dove in and swam back into the tube and brought me back.

After getting me out Mom pinched my nose and blew into my mouth and then pumped on me until the water came out of me. After that I would never let Pop even get close to me if he had been drinking.

Soon the summer was over and it was time for me to start school. The leaves had changed colors and were starting to fall off. Also, the rains that came once in a while were not as warm. I could hardly wait because Mom said I didn't have to wear bib pants anymore. Mom said she would buy me a couple pairs of Levi's for school and make me a regular shirt.

Sure enough Mom made me two new shirts out of flour sacks with print on them. She also had bought me two pair of Levi's just like Pop wore. I also got a belt with a big cowboy belt buckle to hold my pants up. Mom told me I had better try it all on to make sure it fit like it should. After trying it on and Mom checking for a good fit, she said I better get to bed and get to sleep so I would be ready for the day at school.

Like all kids, I tried to talk her into letting me stay up a little longer. Mom said no because it was already past my bedtime. After several minutes, which seemed like hours, I finally started drifting off to sleep.

Morning came soon, with the neighbor's rooster crowing and Mom shaking Lois and I to get us up for school. I was up first and hurrying to get into my new clothes, excited about them, and finally getting to go to school. Soon Mom had breakfast ready and then Dad and Mom were loading us up in the truck to go to school.

After getting there, Mom took me to my room to meet my teacher. Then she told me I would have to wait for Lois and Lobo outside my room before walking home. Soon the teacher introduced herself and also had us introduce ourselves to each other. After all the introductions, we were given pencils and paper and the teacher said we were to keep them in our desk, for our use only.

Then a siren blew and Mrs. Parker said we would be having what was called a fire drill. We then stood up and filed outside, in the same order as we were setting. After a few moments of silence, another siren blew and we were told to return back to our room. When we got back to our desk, we were told this would be practiced every once in a while.

Soon a bell rang and Mrs. Parker said it was time for recess. When school was out I went outside to wait for Lois and Lobo. Lobo was already waiting outside my classroom. Lois then came and we started home. We headed down the dusty road, with a canal running along the side of it. Before long we could see the store ahead of us where I used to meet Lois. We started running until we were finally home. Mom made me change clothes before going out to play.

The next few weeks went by fast because of all the new things happening at school. The whole school year was going by fast and spring time was near. Spring brought its usual warm flash showers along with the pretty green fields and birds singing in the blossomed trees. Along with the spring, I was getting anxious for school to close for the summer vacation.

This really showed because during recess Dick and I were chasing the girls and, if you caught one, you got to kiss her. I caught little Suzy and kissed her, but Dick, her boyfriend, didn't like that a bit.

Dick and I started pushing each other around until we were in a wrestling match. The playground teacher came running and yelling to break it up. Mrs. Parker then reached down to pick me up off Dick because I had a handful of dirt and gravel, rubbing it in his eyes.

Just as she pulled me off him, I came up with a handful of dirt and threw it in her face. Her eyes full of dirt, she dropped me, and as soon as she dropped me, I set out on a dead run. I headed for home down and across a field, then down along an irrigation canal until I reached the old country store.

After reaching the store I slowed up because I saw the principal's car — Mr. Graves was his name — pull in our driveway. I would run a ways and then dive off the side of the road every time I saw a car coming from the direction of the house.

Getting just about home, I saw Mr. Graves backing out of the driveway. I then hit the high grass alongside the road and laid flat until he went past. After some time had passed, I got up and looked to make sure he was out of sight.

Then I started for the house on a walk trying to figure out what I was going to tell Mom. I knew it would be twice as hard on me if I told a lie, so I decided I better tell her the truth.

When I got to the house, Mom saw me and came outside. She asked me what I was doing home so early, I shrugged my shoulders and said I didn't know. That must of made her mad because she sternly said, "I thought I told you to always wait for Lois and Lobo." Then she said I better tell her why I was home early and why my shirt and pants were all dirty.

I started confessing. I told her I got into a fight at school with a boy and threw dirt into the teacher's face when she pulled me off him, and when she dropped me, I got scared and ran.

Mom reached over and picked up a switch and started switching me. At the same time, she said I better hit the road back to school. I thought she would never stop hitting me with that switch.

Finally, we reached the school and I knew I was going to have to see the principal and have a talk with him. Also I knew I was going to have to go apologize to Mrs. Parker and to Dick, the boy I got into the fight with. That was the hard part because I had to do it in front of the whole class.

After that, my mother left after telling me I better be good or I would get it again if I got into anymore trouble.

The rest of the day just seemed to drag on; I thought it would never be time to go home. Finally, the bell rang and I hurried to meet Lois and Lobo.

As Lois, Lobo and I headed down that old dusty road, I couldn't help but think about how I was going to get it again from Dad when he got home. It was bothering me so the closer I got to home, the slower I would walk. Lois finally said I better hurry up or she was going to leave me. Reluctantly, I hurried up so Lois wouldn't get me in anymore trouble than I already was.

When we reached the house, I hated the thought of going in but I did, so I could hurry and get my clothes changed since Dad wasn't home yet. I was outside playing and had almost forgotten about the trouble I had gotten into earlier, when I heard Dad's truck pulling into the driveway. He went into the house and before long out he came yelling to the top of his lungs for me to come and talk with him.

Reluctantly, I answered and went right to him. I could tell by the tone of his voice that I was in for it again. He said, "I hear you got into trouble today and I want to know what happened."

I told him the whole story. When I was finished, he said, "Son, it's all right to protect yourself, but you shouldn't have thrown dirt in the teacher's face and then run home."

Dad said, "I better never hear of you doing anything like that again."

He said he was going to have to give me a swat with

his belt for what I had done to the teacher. After the swat I was also given the old Indian silent treatment. No one would talk to me for several days. Even visitors were told not to talk to me. After the silent treatment was over everything got back to normal. Soon school was out. The weeks had gone by so fast.

3

On to Oregon

The end of school had brought about other changes in our home. Dad had decided we were going to move into town, so he sold the place and bought two lots in town. The one lot had an old house on it, which we lived in while Dad started building a new house for Mom on the other lot.

It was a real busy summer for me because I got to go with Pop on his job, when he topped trees and moved houses. I also got to help him at home on our new house. Soon the summer was coming to an end and it was about time for school to start again.

We were excited because we not only got to wear our new clothes, but we were also going to be starting school in town. Lois got a couple of new real pretty dresses and I got a new store-bought cowboy shirt to wear with my new Levi's. I was glad I didn't have to wear bib-overalls anymore because the ones I had from last year were almost wore out.

Mom had made me wear them picking cotton and grapes that summer. She said I could finish wearing them out playing after school. This made me happy because I hadn't seen anyone in town wearing bib-overalls. In the country, all the boys seemed to wear them, but now that we were living in town it was different.

After the first day of school, Lois and I started walking with Lobo just like we did in the country. Lobo would always be waiting for us, only now he had to wait across the street at the park by the school's crosswalk guards. Lois and I would walk with Lobo through the park and on home.

After several months, Mom and Dad decided to move to Roseburg, Oregon. Dad started selling everything, even the old truck. He said we would get another truck in Oregon for Lobo to ride in, but for now we would have to leave Lobo with Uncle George.

Both Lois and I pleaded with him to let us take Lobo, but he said we couldn't; we had to get to Oregon and get a place to live in. Mom said we wouldn't be able to take him on the bus anyway.

It was hard on us, thinking we were leaving Lobo temporarily. What we didn't know was we would never see him again.

That weekend we had a little farewell picnic at our aunt and uncle's house. Soon the picnic was over and we had to say goodbye to everyone. After giving Lobo a big hug and crying, it was time to go and get our clothes packed for the trip. Mom said she wanted to get everything ready the night before so we wouldn't be so rushed in the morning. That night went fast after our long, tiring day at the picnic.

Morning came and my sister and I could hardly wake up, but after awhile we came alive thinking of the excitement of catching the Greyhound bus to Oregon. We barely had time to get up and dressed when I heard Uncle Tim and Aunt Carol drive up in front of the house to take us to the bus depot.

On to Oregon

It didn't take long and we had the suitcases loaded and were ready to start for the depot. In a few minutes we were there and Uncle Tim and Dad were unloading the suitcases so we could check them in at the luggage department. Within a few minutes, our bus arrived and we were saying our final goodbyes. When we heard the call come over the loud speaker to load the northbound bus headed to Oregon.

Sis and I could hardly wait; we had never been on a bus ride before. Dad and Mom got us seats across from each other. Sis sat with Mom and I sat with Dad. Dad and Mom sat on the outside and Sis and I got to sit next to the window. After the bus driver ran a quick check, he pulled out of the depot and headed north.

A few hours went by and the next thing I knew we were making a stop to pick up more people and let some off. The bus driver then announced we would have about thirty-five minutes at this stop.

Dad said we better go grab a quick sandwich and something to drink because it might be almost dark before we stopped again. After eating a hamburger, french fries and drinking a coke, we just had time to make a quick trip to the bathroom before we heard the intercom announce for all northbound passengers to get aboard. We quickly got on the bus, the driver came around to check tickets and we were on our way again.

Before long I was falling asleep. I had been awake since early morning. When I woke up I looked out and we were in some rolling hills. Not too far down the road, Dad said we would be in Redding, California, soon for a short stop. It was a short stop; we barely got off and they were announcing to load again. This time I sat by Mom and Sis sat by Dad. It got dark and there was nothing to see, so I went to sleep.

The next thing we knew it was morning and Dad was waking us to get off in Roseburg. After unloading our luggage we found a cab and Dad asked the driver to take us to a hotel downtown. After a short ride, the driver let us out at a place called Rose Hotel. Dad paid the man and then they unloaded our things in the lobby.

Mom was up at the desk checking us into a room. Dad came over and said we had better get something to eat up the street first. We asked if we could leave our things there until we returned and the clerk said it was fine and to put them in the corner so he could keep an eye on them. After eating some breakfast we went back to the hotel to get our things and the key to our room so we could rest up for the remainder of the day.

Morning came and Lois and I were wide awake with excitement of being in a new town. Dad and Mom were already up and getting dressed so they could go get a paper to look for a house or an apartment to rent. Dad also told Mom to get hold of our aunt and uncle to show us the area and to get us registered in school as soon as we had lunch.

While waiting to eat, Dad called his brother. Within just a few minutes, Aunt Blanch was there to show us around Roseburg and she also said she knew of a place we might be interested in looking at.

Mom told her we would go as soon as we finished eating if she didn't mind waiting. Aunt Blanch said that would be fine, since it was just a couple of blocks away. When we finished eating, we walked down by a big river with a large steel bridge across it near where Aunt Blanch showed us a big, two-story house. After we looked through the house, we all went over to my aunt and uncle's house; it was only two blocks away from the one we had just looked at.

Dad said Lois and I could go out and play for a little while so Mom and him could visit with our aunt and uncle. Before long Dad came out and told us it was time to go back to the hotel and try to call the people about renting the house.

When we got back to the hotel, Lois and I sat in the lobby, waiting for Dad and Mom to finish making their phone calls. They reached the people who owned the house and Dad said they would rent to us. Lois and I were excited to get a new house in town.

Mom said, "Now that we have a home we will have to go tomorrow and register you kids for school."

We wanted to know where the school was. Mom said it was only a few blocks away from where we were going to be living. The day seemed to have gone by so fast, with all the excitement of finding a new home and getting ready to start a new school.

Soon it was time to get ready for bed. First we had to take our baths and lay out our clothes for school. As soon as we had done that, it didn't take us long to hit the bed and be fast asleep; we had a long and exciting day.

Morning came quickly with Mom shaking us, telling us it was time to get ready for school. We had to hurry so we could grab a bite to eat first. Mom kept hurrying us until we were headed out the door and down the hall to the elevator. I always loved riding the elevator; I had never been on one before.

Leaving the hotel, we went to an all-night restaurant for breakfast. After eating, we walked to school, going exactly the route Mom wanted us to walk to and from school each day. When we reached the school we went straight to the office to register in the new school.

As soon as we walked in the door, a man very politely

introduced himself as the principal, Mr. Hansen. Mom told him our names and said we had just moved from California and she would like to register us for school.

The principal gave Mom some papers and as soon as she filled the papers out, he said he would take us to meet our teachers. After meeting our teachers Mom said she was going back to the hotel and reminded us to come home the way she had shown us earlier.

That first day of school seemed to drag on and on, but finally the bell rang and it was time to go home. As soon as the teacher dismissed us, I cut out on a dead run. In a few minutes, I was running in the lobby and over to the elevator. Getting out of the elevator, I ran to the room and burst in to see Mom and Dad waiting for me.

Mom said we were going to move into an apartment instead of the house. They already had the suitcases packed and ready to go. We would have to wait for Sis to get home from school; she had to go longer than I did. In a little while, here came Sis racing through the door and we went down to the lobby to tell them we were leaving and turn in our keys.

We called a cab and as soon as he arrived we headed for the apartment. The apartment was only a few blocks away, but we had to take a cab so we could take all the luggage.

As soon as we got there, Dad paid the cab driver and Mom was putting things where she thought she wanted them in the apartment. After an hour or two Mom had the place ready to live in.

Then Dad thought it was time to go eat. Mom said there was a little store not far off and she could go get some groceries and fix a bite to eat. Lois and I asked if we could walk with her. She said that was fine with her and it would give her a chance to show us the way to school once again.

After a couple of blocks, we were on Stephen's Street headed in the same direction of the Rose Grade School. We reached the store and Mom bought a couple of sacks of groceries and we headed back home to fix something to eat. By the time we finished eating it was time to get ready for the next day of school.

We got our things ready, bathed, and Dad said we all better get ready for bed; he was going to work in the woods and we had to get up for school. We all had a busy day ahead of us.

When I looked out the window the next morning, it was raining hard. Mom yelled to remind us to be sure we both took our coats out of the closet to wear to school. We quickly ate breakfast, washed up and grabbed our coats to head out to the new school.

4

The Family Fades

The weeks seemed to go by so slow, and once we got home from school all we did was listen to Mom and Dad fight. They fought about everything. One day Mom and Dad decided to split up, if all they were going to do was fight.

Mom came to Lois and me and told us she was going to leave and would be gone for some time. We begged her to stay but she said she didn't want to live with Dad anymore, and he didn't want to live with her either. She told us she had better get going before Dad came home and tried to talk her out of going. She hugged us goodbye and hurried out the door.

Several hours went by and Dad finally came home from work. When he came in he asked where Mom was. Lois told him she had left and wasn't coming back. Dad said not to worry about anything; we would get along okay.

The days seemed longer and longer at school and at home since Mom had left us. I got to where I hated to even go to school. If it was so important for me to go then why didn't Mom stay with us to make sure we went.

Before long Dad got another job and we moved to a

logging camp. Since there was only a few weeks of school left, I told Dad I hated the new school, so he didn't make me go. That made me happier; I could run along the small creek and play in the woods all day. Dad said I could play anywhere I wanted to, but to be careful of Little River because it was deep.

Pop didn't have any trouble convincing me of that; I never forgot the time I almost drowned in the canal gate back in California. The days were going by faster and soon it was summer and Lois was home all day. That summer was extra hot.

Dad was always home early from the logging camp and took us swimming in a hole near Little River. After swimming there most of the summer, Sis and I got to be pretty good swimmers. I didn't think I would ever learn, without drowning. I almost drowned twice because Dad would throw me in and usually only came to get me after I went down the third time.

Soon summer was over and the school year was starting again. We were registered in the Glide Grade School, but didn't go there long.

One day Dad suddenly decided to send us to a foster home. He said he just couldn't take care of us and work too. A lady came and picked us up to take us to the house of some people who took care of kids. After being introduced to Mrs. Dexter, she showed us our beds and introduced us to the rest of the kids as they started arriving home from school.

The next day we were enrolled in the Dixonville Grade School. It was just like the other schools; you were treated like the new kid on the block. As the weeks and months slowly went by, I started to get to know some of the other boys and girls.

We were never together as a family again. Mom and Dad eventually moved to separate towns in California where Mom was killed in a train accident in 1958 and Dad died in 1988. After my sister graduated from high school she married and moved away.

The new home was fun and exciting in a lot of ways. They had a ranch with sheep, cows, goats, and horses. We enjoyed riding horses and having lots of things to do. Each of us had our chores which kept us busy, but when they were done, we could always ride horses.

Summer vacation came and we had more chores, but we also had more time to play and ride. Mr. Dexter also had two work horses he did logging with and us younger boys set chokers behind old Smoky and Hardtimes. Dick, one of the older foster boys, would drive Smoky and Hardtimes with a set of reins. He would yell Gee when he wanted the team to go right and Ha for left.

Summer was coming to an end and we knew we would be getting ready for school. Mrs. Dexter called us all together one day and told us that the people at the foster home agency were going to come and move us all to different homes.

I remember the day they came to move us. It was sad because we knew everyone was going to a different place and we would never see them again. The people from the agency came and picked up Dick first; another car came and it was for Lois and me. It seemed like we rode and rode for some time; I really didn't know how long. Finally, we pulled into a place called Sutherlin and then went out of town a ways to the place where the lady said we would be living.

We got to go through the same introductions as we

did at the Dexter's. This time we were introduced to Mr. and Mrs. Daniels. Lois started crying because she was tired of being moved around because our own folks didn't want us. I told her not to worry; I wouldn't let anyone hurt her. Lucky for us, Mrs. Daniels spoke up and told us not to worry; we would be staying with them until we were old enough to go out on our own.

She then showed us our bedrooms. Lois was given a room with Kris and I was given a room with Ken, their boy that was about the same age as me. That night we were told to go to bed at eight o'clock during the week and at nine-thirty on weekends. Soon it was time for bed and Mrs. Daniels told us to get our baths and set out our clothes for church in the morning.

Early the next day we were wakened by Mr. Daniels calling out, "Daylight in the swamp." Ken and I woke up and slowly got out of bed. After getting dressed, we headed down stairs to wash and eat breakfast before church. I soon learned this was a weekend routine living at the Daniel's.

In a few days it was time to enroll in school. The new school was the same as the others. First, the teacher introduced herself and then she had each of us introduce ourselves. Before long, we were back into the routine of everyday school work and recesses. After a few days each kid would find someone he or she liked and they would become friends. The weeks and months passed until the school year was over and we moved onto the next grade.

That summer was lots of fun, and the best part was that there was no more school work and lots of time to be outside. The Daniels didn't have many chores. All we had to do was stack wood, help plant the garden and keep the weeds out of it. We also had lawns to mow. The first chore in the mornings was to make our beds and pick up our rooms.

We thought the summer was going great until it came time to pick beans. Picking beans lasted about a month and then we only had a week or two until school started. That last bit of time was spent getting our school clothes in the local store and a new pair of shoes for the new school year. Then school would start and we would be back in the same old routine.

The next few years of grade school seemed to pass fast and then I found myself in high school. In high school I got involved in several different sports. Soon I found ninety-four pounds was too small for team sports like basketball and football.

One day, about a week after football season started, there were several of us freshman boys standing in the hall around the area where the high school mascot, a big bulldog, was inlaid in the floor. Since it was about time for wrestling to start, I wanted to stop and check out the school's trophy case. The other boys were looking at the basketball trophies since that was what they were interested in and what they were good at.

About that time, along came Ray, the wrestling coach. He saw Bill, Tom, Earl, Jim, Mark and me standing around the bulldog. Ray then stopped to try and recruit some more wrestlers for his team. Ray started around the bulldog asking each of the boys there if they wanted to go out for wrestling, but when he came to me, he looked at me and didn't even stop to ask me.

I guess he thought I wouldn't be much help since I was five feet, six inches tall and only weighed ninety-four pounds. I remember thinking I'll show him, because I knew fighting was my game, especially since I would be in my own weight class. I was so sure of myself I started going around school telling everyone I was going out for wrestling and I was going to take the state championship.

Wrestling started and it was super, rough training practices. I thought that wrestling three to four matches a night, along with a half an hour of drills, was going to kill me. As if that wasn't enough, after we were finished with that, the coach took us out to run wind sprints against each other.

That first night after practice he told us that was only a light workout, so the pantywaists would drop out. He wasn't blowing smoke out his ears either; the practices got tougher and tougher until we were all in great shape.

The coach was so rough that you didn't dare try to take off even if you had a hurt of some kind. If it were just a mat burn or cold you would go ahead and practice because if you didn't, Ray would have you get up in front of the team and tell them just what kind of ouchie you had so you could get out of practice. You had to make damn sure you had a real good reason for skipping practice.

He also told us we had to run at least five miles each morning on our own. He always said that if we wanted to be a champ we had to eat, drink and sleep it. He gave us eating charts that gave us just certain foods we could eat. This was really rough for me because I loved junk foods and all those were left out of the list. After handing out the charts, Ray told each wrestler what moves he wanted him to practice.

After about two weeks of practice, the first dual match was scheduled. I was really nervous that first night and soon match time rolled around. Sutherlin wrestlers lined up on the mat, starting with the lightest wrestlers on up the weights to heavyweights. Right behind us came the other school and it lined up straight across from us.

The announcer came on the speaker system and introduced each man by his weight. Then each wrestler crossed the mat and shook hands with his opponent. Soon the announcer said, after the introduction of wrestlers, it was time for the lightweights to get ready.

This was the final countdown before the match. I started doing some jump rope warm-ups and then some sit-ups, until I could feel the drops of sweat under my arms. It was caused mostly from the nerves and adrenaline, because of the tension in the atmosphere around the mats.

The next thing I knew the announcer was calling for the two ninety-eight-pound home team and visitor's team on the varsity squad to report to Mat One to wrestle.

The adrenaline really started pumping then because I had never performed in front of a group of people. Also, the opponent looked like he might be able to turn me every which way but loose. Soon, the referee said to shake hands and come out wrestling.

All the tension was over and I only had one thing in the back of my mind and that was to get the other guy down and pin him quick.

Thoughts of the neighbor kid my Dad made me go get to fight again flashed through my mind. Also, how he told me it wasn't size, but speed and leverage that made the difference, and to never stop thinking ahead of your opponent.

Soon I had the kid down and pinned. This seemed to be the way things went most of that year. I only lost one match and that was because I went out and broke training by drinking a couple of root beer floats before the match. I was so sluggish I wasn't able to move, but I only lost by one point.

This taught me a good lesson, for coach Ray's words

kept haunting me all the way home that night. If you want to be a winner, you have to act like one. He always drilled us in practice and told us that you only get out of sports what you put into it.

Ray was always reminding us to get into better shape than our opponents. Even if he was good at his moves you could whip him by being in better shape, but if you cheated in your work-out, you would only hurt yourself.

That night of my first loss his words proved to be true. The rest of the year, I trained even harder and it proved again to be the truth because I went on to wrestle in the state finals. After two days of wrestling, I had won every match so I was going to be in the final match in my weight class.

The big night came and, just before the matches, all the final contestants lined up and shook the Governor of Oregon's hand and then you turned to shake the man's hand you were going to wrestle to determine who was the best in the state.

When I shook the man's hand I was going to wrestle, he took one look at me and said, "I guess I got state cinched."

I thought to myself, "You might, but I'm going to turn you every which way but loose before you do."

The announcement came over the intercom for the ninety-eight pound divisions to get on deck. I knew this gave me about ten to fifteen minutes to warm up. I started doing some curl-ups, push-ups and, last, some jump rope exercises to loosen up.

Then the final call for the Sutherlin ninety-eight-pounder and the Philomath ninety-eight-pounder to report to Mat One.

Our coach came up and told me I should go out there

and pin him because if I did our school would win the most- pinning trophy. That would tie us with Canby, but if we have the fastest pin, we would get the trophy.

I didn't realize it at that time, but he knew I could because I had twenty-eight pins out of thirty matches. I didn't say anything, but I thought, "Okay, it's all over."

The referee met us in the center of the mat and gave us our instructions and then said to shake hands and come out wrestling.

We crossed the mat and shook hands and I turned and took him off his feet and brought him down in a pinning hold, but he landed close to the edge of the mat and the referee gave me two points for a take down and three points for a near fall.

The referee brought us back to the center of the mat in the up position again. This time I took him down again in another pinning position next to the edge of the mat, but he was able to get his shoulder off the edge once again.

This went on until I finally took him down for the fourth time in the middle of the mat where he couldn't get his shoulders off. I pinned him in one minute and thirty-six seconds of the first round.

As soon as I pinned him and started off the mat, the rest of the team came and picked me up and packed me off above their heads. Ray never said much, but smiled and said, "Good Job."

The next three years I took state one more time, but I should've taken it every year. I whipped everyone I wrestled before the state matches, two out of three matches, but lost two years in the semifinals. After losing I gave up even trying because I knew I should have won.

5

Logging Alaska Style

The summer after school got out, I started working in the woods as a logger. Soon I was going to the taverns after work to have a few drinks and dance with the girls. Being small and able to handle myself, I was always in some kind of street fight.

The first time, I was at a logging show where they had log rolling, speed climbing, tree topping, axe throwing, power-saw bucking and hand bucking. That night they had a dance where people came from all over to drink and party.

I went in the dance to check the girl situation out. Lance, a friend of mine, was dancing with a girl and some other man cut in. As he did, they had words and the man who cut in shoved Lance on his butt.

Lance told the guy to help him up and when the guy wouldn't do it, Lance invited him out of town to fight. As they came to the door, they saw me and asked me if I would go along with several others to make sure it was one on one.

I told them I sure would because it looked like several car loads of the other man's backers were going out of town with him. We headed out west of town to a big log pond where floated logs were kept for storage.

When we got there we all got out of the cars and Lance and the man got with it, but every time Lance knocked the guy down, his friends would grab Lance and hold him until he got up. When he knocked Lance down they just let him keep fighting until Lance would get on top of him.

Then they would pull Lance off again. This went on a couple of more times and then I stepped in and said that would be enough of that.

At first, they didn't pay much attention to me until I walked up and took some clubs they were carrying and asked them what they were planning to do with them. The next thing I knew, someone grabbed me around the throat from behind. I threw him on his back and had one knee in his crotch and his throat pinched off with my hand. After a couple of whacks across the nose with my forearm, he was done.

As soon as I got up I noticed at least a hundred people had showed up; they just seemed to be showing up from everywhere.

Before I realized it some coward tackled me from behind and we rolled down over a bank on the roadside, away from the pond. When we stopped rolling, he was on top of me and choking me and I felt like I was about ready to pass out when I buried two fingers in his eyes and he screamed and let go of my throat.

We both came to the standing position and I started thumping him real fast with my feet, knees, elbows and the side of my hands.

Afterwards, he realized he was no match for me street fighting and he grabbed two big rocks and was going to try and use them on me but Ted, who had come along with us that night, came running up from his right side and yelled, "I'll get you, chicken shit, if you try to hit him with those rocks."

He dropped the rocks and took off on a run, but he was too late. Ted was already running full speed and as the coward looked over his shoulder, Ted hit him with his fist while still running full speed.

He hit him so hard, he knocked him back in the ditch alongside the road, and he was out cold.

When Ted and I got back up to the log dump where Lance was fighting, he was just finishing his man. Then cop cars came from everywhere so everybody started running for their cars.

About half the cars were already blocked in by cops so they got names of about fifty peoples who couldn't get away. The cops had blocked the north and south roads leading to the pond. We were lucky; we were one of the many who parked on one of the side roads they didn't block off.

Since no one was hurt seriously, everyone ended up back at the dance, but this time the police had it patrolled. The rest of the night went well at the dance, which was playing my favorite fifties and sixties hits like "Blue Suede Shoes," "Heart Break Hotel," "Twist and Shout" and many more.

After working around Sutherlin a little while longer, in March of 1962 I decided to head north to Alaska. Everything was still untamed and wild. They had lots of logging camps but they had been mostly done away with in the lower states.

Because I liked living in the camps, I decided this was the place for me. I never did like doing my own cooking and house cleaning. Also, it was the last frontier.

Soon the day came for me to catch a plane, so about five drinking logger friends of mine decided they should take me to Seattle to catch the plane. They were really looking for a good excuse to have a party and maybe meet up with some good-looking woman by chance.

The trip to Seattle went fast because everyone was having a good time. We gave a carload of six gorgeous women a cat and mouse chase down the freeway. Everyone was really boozing it up, except the driver. We always had a rule that the driver had to stay sober, so we would get there in one piece. Our driver wasn't allowed to drink at all.

When we reached the Seattle-Tacoma Airport, everyone helped me unload my luggage and get it checked in. When my luggage and tickets were all taken care of, I found my boarding gate and then we went in and said our final farewells in the lounge.

After having a few jokes and laughs about being stranded in Alaska, it was time to board that big Alaskan jet. After we boarded, the plane taxied down the runway and then climbed to its leveling elevation for the cruise to Ketchikan.

6

Northern Adventure

The plane's captain came on the microphone and pointed out all the different landmarks as we were flying over them. This made the trip seem to go fast. Before long, he was announcing we would be making our landing in about ten to fifteen minutes. We would be landing at Annett Island and those going to Ketchikan would be catching the "goose" (a plane) over to the Ketchikan terminal. Those going on to the interior would stay aboard and would depart in just a few minutes.

When I first got off the plane I was surprised to see we were on a flat island. I was expecting to see lots of mountains with lots of snow peaks. Later I found out this was the only flat place they had to build a runway for a big jet to land. Soon the gooses arrived to shuttle the people to their southeastern destinations, like Ketchikan, Wrangell, Sitka, and Juneau.

We climbed aboard the gooses and took off for Ketchikan. We were in the air around five minutes and then we were landing in the water and then taxiing up to a dock in front of the Alaska air terminal.

We went into the terminal where I got my luggage and headed down the street, a little hung over and broke down from the last few days in the bars at home and the trip to Seattle-Tacoma Airport. Then there was the plane trip here. In fact, I was sick, sorry and broke.

The first place I went was known as a logger's bar. I went in and asked if any of the logging camps were needing any men. One of the old-timers sitting at a booth said you can take your pick kid if you know anything.

I told him, "I didn't know much, but I intend to be the best damn logger in the country before many years because I came up to Alaska to learn all the different ways of cable logging."

He introduced himself as Pokey. I told Pokey the next five or six years would be my college in logging. Then I might have the basics down where I could get a job in about any camp I wanted to.

I left the bar and headed down the street. A logger came around the corner and stopped and asked me where I was headed with my suitcases. I told him I was going to get a room until I could find a job at one of the logging camps.

He told me his name was Luke and he was a logger for Thorne Bay Logging and I could probably get on there. "Come along with me," he said, "and we'll go see what the Big Swede thinks of the situation." He was just headed out there anyway to see what time he wanted to catch the plane back, so we went to see the big Swede.

When Luke reached his room, he knocked and sure enough, the Big Swede was in. "Come on in," he yelled out. When we stepped in he said, "What you need, Luke, and who you got with you?"

Luke told him he just wanted to know when he was going back to camp and "if they need some help out there, because this kid needs a job."

In his big, rough voice, Big Swede asked, "What do you do, boy, and are you scared of work?"

I told him work was my favorite pastime and I run most men to death, because most of them can't keep up with me.

He said, "You're sure full of spunk; that's the kind we need up in the big north country." After that, he told me to go fill out an application and catch the plane out in the morning.

Then the hard part came; I had to tell him I didn't have enough cash to catch a plane.

He said, "Hell, that's no problem. Most of the loggers come into town, spend every dime they have and have to charge it to the company to get back to camp."

Luke decided I should stay at his house with his three partners, so I could get up and go out with one of the guys who worked at the mill, where I had to make out employment papers, and then get back and catch a plane to the camp.

Luke told Swede he would see him at the bush pilots hanger at 6:30 so they could get back in time to catch the crummies (busses or trucks) to work.

Luke then looked at me and said, "We had better get going, boy. You did the right thing getting out of that cutthroat country down south."

We started heading back to Luke and his partner's house. We were walking up the hill and Luke said two of the guys worked for the phone company, one at the mill. Luke worked at Thorne inlet.

He said they had plenty of room, not only that but if they were off work, they were usually out on the town with their women.

When we reached the house I found he pretty much

was right about the situation because one guy was going out that night and the other was also getting ready to go. The one that was getting ready to go said to give me his room because he probably wouldn't be home all night anyway. It seems he had a girlfriend he usually spent the weekends with.

Luke asked me if I wanted the room or if I just wanted to stay on the couch. I told him the couch would be just fine. After getting everything straightened away, he said just to make myself at home and showed me where they kept a spare key to get in and then said he had some things to do.

Luke headed out the door. He turned and said, "See you at 6:00 in the morning. Be ready and you can ride to the mill with me on the same bus I have to catch. I thanked him for the hospitality and said that would be fine. I decided to go look the town over a little bit, so I sat my suitcases in the corner and headed out the door myself.

I walked down the boardwalk, then out past the tunnel and past the different bush pilot hangers. As I walked along, I couldn't help but notice how exciting the place was with all the boats and planes there.

After some time I headed back to Luke's place. That's when I noticed it was still daylight outside. In Oregon it would be dark by this time. When I reached the house I was the only one there so I decided to sack out since I was so tired from the long ride to Seattle-Tacoma Airport and then the plane rides to Ketchikan.

Even with the excitement of the plane rides and the new town and friends, I was sound asleep before long and the next thing I knew, I was being wakened by Luke to go to the mill. I splashed some cold water in my face to fully wake up. Luke fixed some breakfast and then we were out the door.

I thanked Luke again for taking me in and we headed down to the corner to catch the cab out to the bush pilots' hangers. From there, I could ride the same cab out to the mill. I arrived at the office, filled out the application papers and the man hiring gave me the address where I could go get a physical. He also told me which bush pilot hanger to be at, and at what time, to go out to the camp.

After the physical was over, I was on my way to catch the bush plane to go out to the logging camp. I checked in at the hangar and sat down by a window overlooking the bay. The fishing boats and cruise ships seemed to be coming and going constantly, and now and then a plane would take off and then another would be landing.

Soon a lady's voice came on the intercom and said all passengers going to Thorne Inlet would be boarding for take off in five minutes. Pretty soon she came on again and it was time to go to the loading dock.

There were five people, besides myself, who headed down the ramp to the dock where the float plane was waiting. When I reached the dock, the dock hand for the airlines told us to go ahead and load. I got on first so I sat up front with the pilot and the rest of the passengers sat in seats behind the pilot and co-pilot's seats.

The pilot got in and started both engines. After running them for a little bit, he motioned for the dock hands to let us go. In a few minutes, we were taxiing down the strait. After the pilot got his clearance, he pulled back on the throttle levers until the plane was climbing out of the water into the pretty, blue sky above.

Several minutes later, we were flying into Thorne Inlet where I could see homes, trailers and several rows of bunk houses. Then, the pilot was setting the plane down on the water and taxiing across the bay to a dock where a man was waiting with a crummy. When we reached the dock, a man was waiting to tie up the plane until we got everything unloaded.

The man in the crummy introduced himself as John, the bull cook. He told us to wait until he got back and he would take us over to another dock, where an L.S.T., a large former Navy landing ship, was located. This is where we got rooms until they could move us up to a bunkhouse on the boardwalk. When he got back, he showed us to our rooms and also gave us a tour of the ship. John told us we could go up to the cookhouse about 5:30 p.m. That was when they started serving.

After putting my clothes away I started my own tour around the ship. I found it had a big kitchen and mess hall. After looking around a bit longer I decided to go into the recreation room, where they had a pool table. After shooting a few quick games, I headed up to the main camp to look things over. While looking around I found the cookhouse, washroom, and commissary.

Then I walked down the beach to the L.S.T to wait for the chow bell. The ship was our bunkhouse and temporary quarters. In about an hour, the bell rang and I headed for the cookhouse.

One of the women waiting tables, seated me with seven older loggers. I sat down and the old guys started to inform me of how things were done at their table. First, they said to ask for things and not to reach, and there was a proper way to ask for things.

When they were finished, I told them they had nothing to worry about. That was the way I was raised to do things. I don't believe in acting like a pig at the dinner table. I didn't realize it, but I guess there were some awful animals that hit these camps and didn't know the meaning of the word manners. After eating I walked back down to the L.S.T to get my work clothes laid out and shoot the bull with the other new men. Along about eight o'clock, I decided to hit the sack so I wouldn't have any trouble getting up in the morning.

I was so exhausted from all the excitement, I didn't have any problem falling to sleep. The morning chow bell seemed to come quick but because of my new job, I was out of bed before I really woke up. After going to the bathroom and splashing cold water in my face, I got my head together and dressed and headed for the cookhouse.

When I finished eating, I went to the office where all the new men were told to check in. One of the side-rods, who told you what logging site, or side, you would be working, told everyone except me where they would be going. I stood there for some time when finally an old hook tender came up and asked me which side I was going on. I told him the side-rod hadn't told me yet. He introduced himself as Spurge and said he needed a man and I looked like I would make him a good choker man.

Spurge went into the office for a few minutes and came back saying to go with him. After we loaded into the crummy, he said I would be working on his side.

Time seemed to fly working on his side but after a few weeks I told Spurge I'd like to learn to climb. He said I would have to go to work on the rig-up and I'd better really want to learn to climb because guys in rig-up worked their asses off.

That night, Spurge came over to my bunkhouse and

said if I still wanted to, I could go to work on the rig-up in the morning. I was going to be working with the Bellowing Dutchman. He also told me that if I ever wanted to I could come back on his side because he always needed a good choker man.

The next morning I went to the office so the side-rod could show me where to go to work. The side-rod told me to wait until the big, Bellowing Dutchman came in and he would introduce me to him. I stood there for about ten minutes before the Dutchman finally showed up.

Barney, the side-rod, motioned me over to where he and some others were standing and introduced me. Barney said, "Dutch, this is George and George this is the rig-up hooker, Dutch." Dutch stood there for a moment looking a little bewildered, and then he turned to the side-rod, snarled, and under his breath whispered, "Hell, they call that a logger."

I didn't say anything, but I thought to myself, I'll show you I can work your ass off you big overgrown, red-haired ape. The Dutchman motioned for me to come with him and he took me down to the line truck, which was our crummy. The rest of the rig-up crew was already there and waiting.

The Dutchman quickly told me their names and jobs and then he told us to get on down the road. Off we went. I was anxious to get to my first day's work on the rig-up. I broke in the hard way, because we were rigging a tree first thing that day.

It seemed like I pulled haywire (cable) around the world at least twice. After finally getting the tree raised, the Dutchman told me to get my climbing gear and spurs on so I could go up and help hang the top guide lines, bull block and haul-back block with Stan, one of the regular climbers. That night I understood why old Spurge said the rig up was a lot harder work than setting chokers. I was so damn tired that afternoon that I thought the work day would never come to an end.

When I got in that evening, it didn't take me long to get showered and eat. After eating, I felt like I could almost put in another ten hours. Going straight to my room, it didn't take me long to find out it was false energy. I laid on my bunk and fell right off to sleep. After sleeping for about an hour I woke up and managed to crawl in the sack. Morning came quick, but I was completely refreshed after a good night's sleep. After eating breakfast, I was ready for another new day on the rig-up crew.

The next few months seemed to go by fast. I was learning so many new things on the rig-up crew. I learned to top trees, rig standing and raised trees, and also how to strip the old trees which had been used on a logging site but were no longer needed.

Another job of the rig-up crew was to move yarders up on the mountain to a place where they could yard (pull) logs to a flat, called a cold deck. Later they would be swung down by another yarder to a place where they could be loaded on logging trucks. When we had moved the yarders back down the hill, then our job was done until it was needed again for another cold deck site.

Most of the time, I was busy helping rig and tear down old spar trees. After about six months of this I decided to move on to another camp. Without giving it a second thought, one morning I got up and headed straight to the office and asked them to call me a plane for Ketchikan and make out my check. They evidently were used to this, because of all the tramp loggers. They said there would be a plane in about an hour and they would have my check ready, along with my ticket.

That didn't leave me much time, so I quickly ran back to the bunkhouse and threw all my clothes into my suitcases and got them down to the bull cook's crummy. He would take them down to the dock before the plane came in. I made a quick run to see old John, the Swede. John was just getting back from breakfast, so I timed it just right. I told him I came to say good-bye because I had just pulled the pin. John said he sure hated to see me go but a man had to do what he had to do. We shook hands and wished each other the best of luck. Then I told John I better get down and pick up my plane ticket and last check.

I headed out the door in the wind and rain. It was raining and blowing so hard that day, I couldn't help but think, "Boy, you sure picked a fine day to pull the pin."

Arriving at the office, I quickly ducked in out of the torrents of rain driven by big gusts of wind. The head man, Don, told me the plane would in at seven o'clock and I could catch it to Ketchikan. I looked at my watch and saw that I still had fifteen minutes, so I decided to go have a quick cup of coffee. As soon as I gulped my coffee down I headed back to the dock, just in time. The old goose was just landing.

When I arrived at the dock, the bull cook was unloading the mail into the crummy. I quickly grabbed my suitcases and climbed aboard for the stormy ride to town. The pilot told me to come sit up front with him I climbed into the co-pilots seat and then the pilot started the engines and the bull cook shoved us away from the dock.

Within seconds, we were headed down the bay and into the air. The ride was really rough when we hit the straits because the winds hit harder in that area. I was glad to be in a goose because they were a two-engine plane that could fly on one engine and could land on land or water.

After flying over Guard Island, it was just a few minutes and we were making our approach for water landing and then setting down in front of the terminal. I gathered my luggage and went straight for the terminal to flag down a cab. Flagging a cab didn't take long at all. I set my luggage in the back seat and then jumped in the front and asked the driver to take me to a hotel downtown. The driver asked which hotel I wanted to go to and I quickly responded, "It doesn't matter even if it is in the main part of town."

In just a few minutes we were pulling up in front of the Ingersoll Hotel. I got out and paid the driver. He quickly made change for me. I grabbed my bags and headed into the hotel desk and asked for a room. After getting a room and unloading my luggage, I headed for the old Saw Dust Pit bar to celebrate. The band wasn't in yet, so I had a quick drink and headed up the street to the Fo'c's'le bar, where all the tramp loggers hung out.

As I was walking up the street, I thought once again how nice it was to be in town instead of being up in a tree hanging the rigging with the winds blowing sixty to eighty miles an hour. As I reached the Fo'c's'le, I ducked in dripping wet from the torrents of wind-driven rains. I walked up to the bar on the end, ordered another seven-seven and just sat there sipping it and enjoying my freedom of being out on the town.

I started contemplating going north and finding a job. I was about to order another drink when Don, a young man I'd seen in camp came in and sat down across the corner of the bar from me. I asked him if he'd been working at Thorne Bay. He said he had been there but quit a week ago.

We introduced ourselves and just sat there shooting the breeze. During the conversation, I mentioned to Don that I was thinking about going up north to work and see what the country was like. Don said he had been asked to leave town, but hadn't decided where to go for sure. I asked him what the problem was. He said he had a drinking and driving problem and the police told him that if he would leave town, they would lower the charges since no one had been injured.

I told him I was going to catch the Alaska ferry north tomorrow at ten o'clock and he said he would get a ticket and meet me on the ferry. We would look for another logging job, up out of Juneau or Sitka. Don said he had to hurry and get his ticket and take care of some business, so he took off.

I decided if I was going to leave, I would boogy that night and maybe try to pick me up a woman. After having a couple of more drinks, I thought I'd better get a bite to eat or I'd be too drunk to chase women.

Later that night I met a young lady named Janet. Janet liked boogy and so we seemed to hit it off right from the very beginning. We went from bar to bar drinking and dancing. Early, about dawn, we both decided it was time to call it a night since we both had things to do. After breakfast, I called a cab and took Janet home. I returned to my room at the hotel, jumped in bed and was sound asleep within minutes.

All too soon, the buzzer was ringing and I knew the desk was calling me to get up. It seemed like I'd hardly had any sleep. When I stood up, I had a giant-sized headache from too much booze and too little sleep. I hurried to shower and pack my things, called a cab, and went downstairs to turn in my key. Soon the cab driver came in and helped me load my things.

I told him to head for the ferry dock. When I got to the dock, Don was already there getting ready to board. I hurriedly paid the driver, unloaded my things and went to check in. Don said he was wondering if I was going to make it or not. I told him I was running it close because of my long night out with Janet.

We put our luggage away and started walking around to check out the ship. I told Don I was going to have a drink and try to soothe my hangover. He said he could use one too, as he had a bit of a hangover himself. We found the bar and was disappointed because we found out the ship had a ten o'clock bar time. With a half-ass laugh, I looked at Don and said I guess we'll just have to suffer for a few more hours.

Walking around with the shakes, it seemed like forever before the bar finally opened again. We sat down and I ordered a seven-seven. I started drinking it fast and before long, I was feeling better. By the time, I reached the third drink, I was beginning to feel normal again. I was sitting where I was going to spend the rest of the trip.

Soon we were pulling into Sitka for a short layover. As soon as we were loaded, the ship was on its way again. I could hardly wait to get going. The bar was always closed before docking and while docked. The bar opened after the boat pulled out. I remember thinking why don't those damned people ride another ship so I can get on with my drinking.

It seemed like a short time and the bar was closing again. It didn't bother me too much this time, as I was already pretty well pickled. When they called last drink I was already on my way up forward to catch some shut-eye. It didn't take me long to fall asleep. The creaking and rolling of the ship along with the waves spraying up over the bow just seemed to put me to sleep faster.

Morning came and we were only about an hour from Juneau. As soon as we pulled into the dock, Don and I both were headed for the Red Dog Saloon. We quickly checked the job situation and decided we would grab a plane to Sitka. It sounded as if jobs would probably be better for loggers there. The flight wasn't long.

Landing at Sitka on the water, the P-V-Y turned up on a cement turnaround pad. Then the pilot and the co-pilot came back and opened the door. Don and I grabbed our bags and headed for a place to stay. Since we were starting to get a bit low on funds, we told the cab driver to take us to a cheap, clean place. He said he knew of a place. Don and I looked at each other and laughed because we knew how much money we had blown on booze. The cab driver kind of laughed and said, "You boys don't have to worry. I've been there, too and besides there's plenty of work around here for loggers."

Lucky for us, when he dropped us off, we were straight across from a working man's restaurant. There was this old run down hotel and we just looked at each other. The cab driver knew we were having second thoughts but didn't really have much of choice. We paid the driver and went in to check on a room. We told the clerk we wanted a room with two beds and a good view if possible. He said luck was with us because he only had one left. We checked in and went to put our luggage away.

We couldn't wait to go check this town over. The sun was gorgeous that time of day and we enjoyed the walk.

Before we knew it we were standing in front of a bar and we thought it was just as good a place as any to check if there were any of the camps needing help. This went on for about a week and a half. By this time, we were really in need of cash and we finally came across a lead on a job.

An old logger told us he'd just quit a camp across from old Sitka. He also told us they would be needing two men because his partner had quit too. Don and I didn't waste any time at all getting an appointment to meet with the camp push. He and his wife were staying in the best apartment complex in town. At least it looked that way to us after staying in the shack we had been staying in.

We were so low on money we had been drinking wine and only eating one meal a day. One young lady felt sorry for us and told us one day to come back in the afternoon and she would buy us lunch. All the goodies in the soup pot would be in the bottom by that time and we could feast on that. It didn't take us long to tell her we would surely be back later.

When we got outside we couldn't help but laugh and wonder how she figured out we were so hungry. Really, it wouldn't have been too hard seeing us day after day walking up and down the streets looking for something to do.

Anyway, maybe things were going to start looking up. We had an appointment to see a man named Harold about two jobs on the largest skidder in the world. Since it was about time to meet him, we walked toward his apartment. We got there just in time, as he had just pulled in. We talked to him for quite some time, and he told us we were hired if we could pass the physical tomorrow. We told him thanks and hurried outside and down the street.

We couldn't help but laugh. It was so good to have

a job again and know that we were going to get three squares again. After having a good laugh, Don said "Let's go get the bottom of the soup bowl."

When we reached the restaurant and walked in, still laughing, the young waitress couldn't help but wonder what was making us so happy. We told her we had finally gotten jobs and it was a good thing because we were both down to our last few cents. She said she already knew that and that's why she had offered to buy us lunch. We ate our bowl of clam chowder and thanked the waitress for taking care of us. We told her we would be back to see her as soon as we came in from camp.

We headed for our room to get things ready and organized for the boat ride to camp early the next day. The rest of the day just seemed to drag by because of waiting to leave for camp, but night came and then morning was there and we were on our way.

We reached the old Sitka dock early and had to wait a few minutes for Harold. Finally, Harold showed up and told us which boat to load our gear in. After putting our gear in the big, eighteen-foot speed boat, we were racing across the ocean toward a camp in a cove on a river. We were excited to be reaching a new job and home.

We pulled up to the dock in Nocquicena Bay. We got our gear and then walked up to the cookhouse. Harold introduced us to the cook, Walt, and his wife, Alice. Alice was also the bull cook and flunky. After showing us to our rooms, Alice asked us if we would like something to eat. It was about time for breakfast to be served and we anxiously accepted her invitation. She then guided us to the chairs where we would be eating while we worked in this camp.

In just a few minutes Alice was ringing the chow bell

and men started rushing in. Soon the place was full and Walt was introducing everyone by complete names. This was unusual, that he knew everyone so personal, but this was a small camp with about eleven crewmen and that was including us.

Harold, the camp push, stayed just long enough for breakfast, and was off again as soon as he introduced us to the side-rod. He told us to get our rigging clothes on and meet everyone down on a boat, the *Sitka Logger,* in about a half an hour, to go over to the skidder. I was really excited about this trip because this was supposed to be the largest and only skidder left in operation. The day went so fast and was full of all kinds of new and exciting things.

After a few days and getting to know everyone, we both decided this would be a great camp. All the men were young and the cook and his wife were known as among the best, if not *the* best in the whole southeastern Alaska.

Soon it was time for the loggers' contest in Sitka. This usually took place during the Fourth of July, so the camp push let Vern, the guy in charge of running the boat, take the whole crew in town for three days. Since I had just had a throwing axe stuck in my knee, I was quite ready for a few days off. Everyone else was looking forward to the break too, since this was also a chance to party and chase women.

When we got to town the first thing I did was go get me some new threads and then start hitting the dance clubs. This trip ended up being nothing more than a three-day drunken party. Since I never could drink much and still keep my composure, I ended up drunk and passed out when the crewman came to get me to go back to camp. I could barely remember them picking me up and throwing me in a taxi back to the *Sitka Logger* where everyone was getting aboard.

We were sailing across the open water for Nocquicena Bay when about half way, I got sick. The crew told me and I even had to ask them to help me out of the hole to the back of the boat where I could heave up some of the alcohol that I had consumed in the last three days.

I faintly remember heaving let alone anything else, but I guess I fell overboard and Vern had to turn the ol' *Logger* around and come back alongside me to haul me back aboard.

The next morning I woke up and asked Don why I was lying on top of the bed with coveralls on. He couldn't believe I couldn't remember what I had done. I couldn't, even when he repeated the whole story. That morning I was not only hung over but feeling quite stupid when we walked into the cook shack Not so much for myself, but mostly for Walt and Alice. I really respected them. Anyway, after a few days of comments and laughs, everything was back to normal.

We finished up the sales in Nocquicena Bay and were getting the camp and skidder set up for a move to Rodman Bay. The move was really exciting because we put the bunkhouse and cookshack on big, log rafts and then laced them all together. A couple of tug boats came out of Sitka and towed us to a new cove where we set up camp. We even stayed on the water hooked together like we were when towed out.

The new logging site was a workhouse. That "largest skidder in the world" took a lot of cable to pull the cut timber down the mountainside. We had to make a complete new layout. That meant packing coils of cable weighing one hundred and twenty to one hundred and fifty pounds to the top of the mountain. The cable would be used to haul the bigger lines up. This all took place after we got the skidder in place.

Being young and tough, we would race at everything we did. It was a mile round-trip from the shore to the top of the mountain where we logged. The logging was done in large sections so the after effects would be like a checkerboard. That way the land where the timber was harvested would replant itself from the section left standing.

After starting the new layout, Vern decided I could race a speed climber we had in camp to the top of the mountain and back. I told him I really didn't want to because I wasn't training and in good shape. Well, I didn't have any choice because he told me he had bet the guy already. I guess I could take a coil up and come back down faster than him.

That day, the guy I was supposed to race and I had already packed four coils and were on our fifth. The other guys were only on their third. Since the guy I was racing didn't smoke or chew, he ended up winning the race by just a few seconds. But just to outdo him, I grabbed another coil while he was lying on the beach soaking in his so-called victory and trying to recover from the race. I ended up packing six, the other guy five and the rest of the crew only four coils that day. So much for fun, now its time to get back to work.

Soon after that we had the skidder operating again and time seemed to fly. It was time for the colleges to start in town. That meant our whistle punk would be quitting. Harold couldn't find anyone to take his place so he asked me if I would do it. I told him, "Maybe for a week, but not any longer" because I was used to working hard and not standing around fighting bugs and tooting whistles on a hand rig, signaling for the logs to be hauled in.

The end of the week came and still no whistle punk was hired. The bug in the battery-operated tooters they had in those days wouldn't work if you got behind even a twig because you didn't have an unobstructed straight line from the bug to the whistle.

It just so happened that the day Harold came out the bug wasn't working. He came up the hill to see if I had the whistle code down. He took the bug and motioned to the operator to stop the rigging while it was empty and said he'd have me send him some whistles to see if I had them fixed in my head. After finding out I knew the whistles, he started back down the hill and the bug began acting up again.

This happened several times and he asked me if I really knew the whistles. I told him, "Hell, anyone could learn the whistles, but you have to have a bug that works."

He decided to repeat the checking procedure just to see if I was being smart-ass or if I knew anything. The same thing happened again and then the crewmen started laughing.

Needless to say, by this time, I was getting a little perturbed and I walked over and called the operator and told him to get me a plane and to tell them I would be flying to the bunkhouse to get my things.

I turned to the young men and reminded them that I had said, "I won't last long on the whistle and as far as I am concerned the side-rod can take that bug and shove it up his ass." I handed the bug to the side-rod and reminded him I was on my way to catch the plane.

When I reached the float and the skidder a plane, that had been out at another camp close by, landed beside the skidder to pick me up. The day was a beautiful day for tramping; the sun was out and the sky was clear; you could see for miles. It was always pretty because of the rough terrain with the water around it, and the snow capped mountains.

Soon the plane was landing next to the floating camp. I told the pilot it would only take me a few minutes and I would be back out with my gear. After running in and packing my things I quickly got back to the plane and loaded my gear. Then I made a run to say goodbye to Walt and Alice and to thank them for everything they'd done for me. Within a few short minutes I was back at the plane and in the front seat alongside the pilot.

After pushing the plane off, the pilot got relocated in his seat and made a quick check. We were headed down the bay and then we were in the air. Flying past the skidder I could see the men working and then the plane was climbing over the ridge. In about thirty minutes we were landing in the water alongside Sitka.

I unloaded my things, paid my plane fare, hailed down a cab and was on my way to Alaska Airlines to catch a ride on the million-dollar crummy in the sky. Upon arrival at the airline I purchased my ticket, checked in my luggage, except for an overnight bag, and started walking down the street to pass time until departure time for the plane to Ketchikan, where I could catch a 727 south to Seattle.

When I finally arrived in Ketchikan, I had a layover until the next day at nine o'clock. I grabbed a cab to downtown where I rented a room for the night. I got my key and headed straight for my room where I could clean up for a last night on the town in Alaska.

I couldn't wait to go check out the bars and the woman situation. It was a booming logging town and was going pretty well for the middle of the week. After checking out a couple of the old hot spots, I noticed several single women sitting at the end of the bar, drinking by themselves in the old sawdust pit, as we called the place they hung out.

I sat down by one of them and ordered a drink and, seeing that she was ready for one too, I ordered two instead. I introduced myself—her name was Lynn—and asked her if it was all right that I'd ordered her another drink. She responded that it was and soon we were talking, and dancing to the music of the band.

After dancing about an hour Lynn told me she had just come up from Seattle and went to work in this place tending bar during the day. She evidently had just got off work and had sat down to have a relaxing drink just before I'd come in. I asked her if she needed to get home right away, and she said she just needed to go home long enough to freshen up a bit, then she would come back.

I figured I'd just been given a good, old-fashioned brush off, but I told her I'd wait for her. She said she would be back as soon as she could because she really felt like dancing and she liked the way I danced. I'm sure she could tell I was ready to party and dance all night. I told her to hurry so we would hit all thirty-two bars and see which have dance floors.

Sure enough, before too long Lynn came and, boy, oh boy, was she ever decked out. I complimented her on how pretty she looked and we sat down to order another drink. After sipping on that drink for awhile, I asked her if she was ready to boogie to "Twist and Shout," an old Chubby Checker hit. She quickly replied, "What are you waiting for?"

With that reply, my "last night" was just beginning. One thing great about Alaska was the nights never ended unless you had to go to work the next day. The bars were open all the time except for one hour. Since Lynn had got a friend to take her place at work the next day, we boogied till the wee hours in the morning.

Before morning came, I realized I was sure glad I'd met her. She was one hell of a dancer and loads of fun. The next morning she went with me to catch the Alaska air shuttle to Annette Island and it was almost like I'd known her for all my life. I sure hated to leave her. After a lengthy goodbye kiss, I said goodbye and told her next spring I'd look her up.

I got on the goose and moved up to the front passenger seat. I was the first passenger. The pilot followed me aboard and told me to move up front with him. I asked him if he thought we would really be flying because it was blowing sixty to eighty mile an hour winds in the straits. He said we sure would be flying.

After all the other south-bound passengers got aboard, the pilot started up both engines, made a quick equipment check, and signaled the dockman to untie us and shove us off. As the plane taxied out the pilot looked over at me and, because of the winds, he asked me to pull back on the wheel with him until we got off the water. Soon we were out in the strait headed down the channel at full throttle.

At first green water was hitting the windshield and then we were on top of the waves, skipping across the white caps, and then into the air. It was a real hair-raising experience as we sailed across the sky alone with the ups and downs from the air pockets.

We were flying sideways because the wind was hitting us in the side about eighty miles an hour. I didn't give it much thought until we were headed for the runway sideways. The turbulent ride brought back old memories of having been lost in a storm and thick fog.

Suddenly the pilot touched down and at the same time snapped the tail around so the plane was straight with the runway. Then he touched the rear wheel down.

7

Mexico: Foes and Friends

We taxied up to the terminal where the 727 for Seattle was already waiting. Going inside to get out of the cold wind and rain, I waited for the boarding call. In about five minutes a call came over the intercom: "Seattle-Tacoma passengers, please start boarding." We hurried out into the weather and quickly went up the stairs to show our tickets and find our seats.

It didn't seem like long before we were approaching the Seattle-Tacoma airport. The first thing I did was run inside to call a friend from Roseburg to see if he could fly up to pick me up in Eugene. Dave said he had no problem with that as long as I paid for the plane rental. I told him that was fine and also told him what time my flight would be leaving Seattle for Eugene.

Upon my arrival in Eugene, I started looking for Dave. Sure enough, he was there waiting so we loaded my things and headed for Roseburg. Soon, we were flying into the Roseburg Airport. I went inside the terminal to pay for the plane and then I was headed for a room downtown where I'd live until I found a place to rent.

After finding a room, I thanked Dave for flying up and getting me. Then he headed for home. I cleaned up and walked across the street to get a newspaper to check on rentals. I looked through the paper and checked off some places, but I decided to wait until the next day to go look at them. It was already starting to get late and I wanted to get something to eat.

After supper I relaxed a few minutes, then I walked around and checked out one of the local dance bars. When I went inside I ran into a couple of old friends who had also just come back from Alaska. Pat and Dan were staying in a local trailer park. We visited for a while and decided we would rent a big house or apartment for all of us and we could share expenses.

The next morning, after a long, rough night of drinking, Pat and Dan woke me up to tell me they had found a house with three big bedrooms. We talked it over and decided to take it.

After a few weeks of living in the house, Dan and I decided to go on down to Guadalajara, Mexico. This is where I had originally planned to go after coming home from Alaska. Dan said he wanted to work one more week and give the logger enough time to get another man. After the week was up we were off for Mexico.

First, we went down through San Francisco's Big Bay Area because Dan wanted to stop and see a friend or two he had known when he was working and living down there a year or two before. From there we headed down to the Nogales where we planned to cross over into Mexico. When we reached Nogales, Arizona, we got our passports.

In about two or three hours, we headed down to Mazatlán, a little town along the coastline. The time went fast as we drove through the desert country. Soon we were driving into Mazatlán. It was early in the morning and there wasn't much doing so Dan and I checked out the beach front.

We just walked up and down, looking at the water in amazement. We wanted to see it up close; it was so blue it almost looked fake. It was even more beautiful up close than it was at a distance. The beach was also whiter than any beach I'd ever seen. You could look into the water and it was such a clear blue you could see the white, sandy bottom.

Another thing I noticed was that the early morning air wasn't even chilled; it seemed to be just a perfect warmth. We spent several hours running up and down the beach and checking out the waterfront stands. Then we moved on to the various night clubs and checked them out. Since the place was already coming alive with people, we decided to hire a local boy to watch the car.

After several drinks, we headed south to Guadalajara. We headed south, with the intention of taking turns driving while one of the other got some sleep. We figured by trading off driving we would get there quicker and rent a room and get cleaned up.

After several hours we reached Guadalajara. Dan was driving but it wasn't long before he pulled over and asked me to take the wheel. He said he had never seen such driving like this, not even in San Francisco. It didn't take me long to figure out what he meant. Everyone was in such a hurry.

As I pulled up to the first red light I got my first taste of their impatience. I was sitting there waiting for the light to turn from red to green. They started honking as the light turned amber and started siga-siga-siga meaning to go before the light turned green. I decided to wait for the green light so I didn't get any tickets in Mexico. After several lights I got to where I didn't even pay any attention to the honkers.

We spotted a motel and pulled in to rent a room. After cleaning up and walking around checking the place out, we decided to rest up for the big night. Later on in the evening, we headed out to see what the big city had to offer.

When we entered the first bar it didn't take us long to figure out that the Mexican people liked to party. We had several drinks and watched bands coming and going, with their dancer doing every three or four sets. Then we were off to find another bar.

As we were entering the second bar, we ran into Sanchez, a young American who spoke Mexican. He would show us around if we would pay for his drinks and girls. We said, "Sure."

He started showing us all the hot spots, including a bull fight parade where they were running the young bulls through the street, along with dancing and shooting off all kinds of fireworks. After walking around for several hours, in all the entertainment, we decided to hit some of the other night clubs and go to the bull fighting events some other time.

We kept going from night club to night club. We ended up in a club full of people drinking and dancing. Within an hour or so we each had got together with a girl we liked and were having a good time.

Everything seemed fine until Dan ordered a round of drinks and the bartender wanted his money. We had both run out except for the twenty I'd just laid on the table. We looked around and couldn't find it on the floor or anyplace.

Pretty soon the girl I was with spoke up and said she

had seen Sanchez pick up the money and put it in his pocket. Dan turned to him and asked him to pay for the drinks. Needless to say, he denied it. He said, "I no take the money." I figured the girls with us weren't lying, so without thinking I got instantly pissed.

I grabbed him by the shirt and jacked him up on his tip toes. I told him he didn't have to steal, I would have given it to him if he needed it that bad. About that time, a Mexican man about thirty or thirty-five ran over and reached to grab my shirt.

Out of the corner of my eye I saw him coming. When he reached me I let go of Sanchez's shirt that I had rolled up under his chin and came around with the back of my forearm and wrist. The blow drove the oncoming man back across the floor.

Dan, being a quick thinker, shoved a chair toward me and was on the floor with his chair within a split second. We had it all planned in advance that if anything like this happened with a large number against us, we would back up to each other, watch the right side and that would cover the other man's left.

Thanks to the two women we were with we didn't have to start breaking heads. They started talking and crying like crazy and then the crowd started to go back to their tables.

Dan and I looked for Sanchez but he evidently got lost when all the crowd gathered. We sat back down at our table and decided we had better go back to our room, where we kept most of our cash, and get some more money.

Now we had to decide who would go after the cash and who would stay with the girls. We tossed a coin and Dan had to go back to the room while I got to stay.

After about an hour Dan came back and we paid for the drinks and decided to take the girls with us. Later we drove around in search of a villa to rent for a month or so. We found one and went back to the motel to check out and get moved into our new home.

That day, Rita and I and Dan and Patsy just laid around and enjoyed the spacious villa with its many rooms and sun deck. Soon night came and we took the girls back home. That night we stayed in the villa, but found out it was way too big for a couple of tramp loggers to feel comfortable in.

The next day we went to see some of the country and drove towards Lake Chapala, not far from Guadalajara. We looked over the shore and started driving back to the villa to spend some time hitting the different clubs. As we were driving we noticed a motel and decided to stop to pick up some postcards and gifts for people back home.

As soon as we entered, we ran into Sanchez, the guy who had showed us around and then stolen our money.

Dan saw him and was ready to run a head on him when I grabbed his arm and said, "He's really not worth it." Dan gave Sanchez a few choice words and put the fear in him, and then we headed for the car.

We spent several more days in Guadalajara and then headed back home. We drove north out of town. All of a sudden we started hearing a squeak in the front wheel. Lucky for me, Dan had worked on cars and knew what to look for. He checked it out and said it was a wheel bearing, but it might last till we reached the good Ol' United States.

We drove for several hours and the squeak kept getting louder and louder until we finally decided to stop and fix it. We checked it out and found that the bearing had gotten so hot it had melted and welded itself to the wheel shaft.

Here we were in the middle of nowhere and we had to decide who would walk back to get a fifty-five Studebaker front wheel bearing. Tossing a coin seemed to be a good way of making decisions, so we tossed a coin and I won. Dan had to go for the wheel bearing.

He hadn't walked far when I saw him flag down a bus and head towards the last village we had passed.

Along after dark I was sitting in the car when a man with a sombrero came along with his son, who was riding a donkey. They stopped and talked with me for awhile. It was hard to understand them since they couldn't speak English, but I had a translation book and we managed to get a few things across.

I could tell they had been working hard in the fields so I gave them some of the pencil and paper for the boy and also some pots and pans I had brought along.

After they went on home, here they came again about two hours later. They had brought me some coffee. They visited awhile longer and I thanked them and they went off towards home again.

I remember thinking of Luke, who helped me get a job at Thorn Bay Logging, and his friends and how there are nice people in all walks of life. Soon I was asleep.

Daylight was starting to break as I woke up. In the distance I could hear a rooster crowing along with the sun trying to come up over the mountains.

Within a half an hour or so, here came Dan with a man who had a wheel bearing for the old 1955 Studebaker. Dan was saying I'd never believe he found it in that last village we passed through. He started describing it as having anything and everything.

Soon the man had the old bearing filed off and had put on the one he'd brought from town. I remember holding my breath, thinking it might not fit, but it fit perfectly.

He put the wheel back on and we paid the man and thanked him for his much appreciated help. Then we were on the road again.

8

Fights, Cops and Frisco

Oregon was a welcome sight; it sure was pretty country with its green mountains and clear skies.

Upon reaching the Roseburg and Sutherlin area, Dan moved home and I moved in with some friends who were living in a three-bedroom apartment.

This was in the early sixties so there were lots of dancing, parties and drinking. In a few weeks I met a girl and we took off for Reno, Nevada, and got married. This only lasted about a year and I decided to leave since we just couldn't get along. I guess it was time to head back to God's country, Ketchikan, Alaska.

I reached Ketchikan once again and spent several years tramping from one logging camp to another. Along the way, I met up with Pooki, Slater Jack, Jessie and Don; they were all old-time tramp loggers. They even put some of them in a movie made from the book *Tramp Logger*.

After the last summer season up there I headed back down south to work for the winter months around Sutherlin, my old hometown.

It was the year 1965 and I went home because I was

thinking of getting out of the logging occupation completely. That summer, Timber Days, an annual event in this town, came around and that made my decision to get out final.

The thing that really put the icing on the cake was when a friend, Doug, who was probably my best friend throughout my school days, —he and I—decided to tie on a three-day party for the Timber Days celebration.

It was almost August and it was really hot, so it seemed like the perfect time for a good beer-drinking party.

On the last day of the celebration we had just come back from a keg party when two cops pulled us over. Doug was driving and I was asleep, or passed out, whatever.

Doug woke me up and told me they had stopped us for the fourth time that day. He was getting a little sore about being stopped so much in one day.

While one cop who was checking Doug, the other cop came up to my side and started asking questions. One thing led to another and then we started having some heated words.

He told me to get out of the car so he could check it out for booze. I told him to go ahead and take the car apart, but put it back like he found it because there wasn't any booze in it, and besides I was old enough to drink if I wanted to.

Needless to say, he thought I was a real smart ass. But I proceeded to tell him that all he was good for was to harass people and that he was a horse's ass. And then I said he wasn't even an ass because that was part of a man.

The cop, named Roy, who was checking Doug, said, "They're clean and he has no record." But the one checking me decided to take me in on some stupid charge.

He started pushing me back towards their rig and Mike, the other cop, was trying to tell him to stop harassing me. He said, "If you'll just talk to him and not try to push him he'll go with you and probably won't give you any trouble."

When he pushed me inside their rig, an Econoline Ford with the motor hood extended into the front seat, I hit my head on the roof of the car, I got pissed and really started yelling some rough, choice names.

Then the cop grabbed my shirt and shoved my head down behind the driver's seat. When he did that, that put me upside down with my butt up on the motor hood in the cab.

That really did set me off. I told him he should have a skirt on because he wasn't even a man.

He reached down to unsnap his clubs and told me I better shut up or he was going to brain me. As soon as I saw what he was doing, I decided there was no way I was going to let him use them on me.

I managed to get my left arm out from under me. I grabbed his club with my left hand and his ear with my right hand.

I wrenched his club out of his hand and started swinging.

Mike's partner, Roy, had just gotten back into the Econoline and I really had no intention of hitting him, but I barely nudged him with my shoulder and knocked him on his rear on the pavement. After all, my fight was really just with Mike.

I quickly jumped back out on the passenger's side and threw the sap out in the field. I turned to Mike and told him, "Come on, now it's time for your education."

The damn fool started towards me and that was his

second mistake. I dumped him on his ass on the pavement and started thumping him a little bit to give him some of his own treatment.

In the meantime, Roy had called for some assistance and, real quick, they had the chief of police, city cops, sheriff and state cops on the spot.

They quickly formed a circle around me, so I just let the cop go. I could clearly see I was out-numbered.

Rick, the chief of police, asked me if he could talk to me. I said that was fine, but he better keep that other S.O.B. away from me. While talking to me, the chief put a set of handcuffs on me.

Everything was all right until that S.O.B. started towards me. I told them to stop him.

The chief tried to shove him back, but he got a little close so I broke the handcuffs and pinned the S.O.B to the pavement.

All of a sudden someone from behind hit me above the right ear with a sap.

When I got hit, I dropped to one knee and came back up swinging with the back of my arm.

I knocked a cop flat. Then I was hit from behind again and was out cold on my feet. I faintly remember someone walking with me to the chief's car. Then they hauled me to jail.

The next morning the chief took me to breakfast. He told me they had booked me and set bail at five hundred dollars. The charge was resisting arrest.

I said, "Hell, if I was resisting arrest, I would have taken that asshole's gun instead of his clubs. I just wanted to give him some of his own treatment. The chief said, "You've got to be one of the toughest men, pound for pound, I've ever seen."

He told me he would reduce the bail to one hundred dollars and let me go if I promised to stay off the booze and post bail.

I agreed, paid the bail and decided it would be better if I left town for awhile.

I moved to Roseburg with four buddies who were also ready to get out of town. At first everything seemed to be going smoothly, and then after about three weeks everything started all over again: the drinking and parties, and women coming and going.

One day an old school friend showed up and we decided to go out on the town and hit all the so-called hot spots. We hit everything from pool table bars to the dance bars. Closing time came and we decided to stop at the Island Cafe for some breakfast.

As soon as we entered the door we saw six men sitting alongside each other at the bar. Being the sixties, they had long hair.

Right away Jason began making smart remarks about their hair. I could see that was going to cause trouble. I tried to tell him it was just a fad.

Jason let it ride for awhile and then started to make more remarks. He told them they were just misfits and that was the way they got attention, and on and on.

Finally, I told him to shut up or he was on his own as far as I was concerned, if he started anything. I guess that made him think, and he decided to stop. I thought we had gotten out of that one easy.

When the six guys got up to leave, the leader was standing to my left, at the cash register. He turned to Jason and asked him if he had a problem with long hair.

That's all it took to get Jason's mouth started again. Once again I asked Jason to shut up, and then I turned to the man who had asked the question and said, "Let's just forget it, there is no problem."

I guess he didn't like that. The dirty son of a bitch turned to me and asked me if I wanted some of it. I could see there was only one way to settle this and that was to kick the shit out of him.

I turned to him and told him to get his ass outside and I'd show him the best thing for him to do is to keep his mouth shut. I had just been trying to smooth things out.

When he started jacking his jaws again, I told him it was just his ass talking, when his mouth knew better.

When he got down to the street he wasn't much, but all of a sudden I was catching it from all sides. His friends had knocked out Jason when he dumped one of them. Another one of the gang kicked him in his head.

Then they came over and started hitting and kicking me from behind, the side and any place they thought they could rush me.

After five to ten minutes of that, someone in the cafe called the cops. The six guys ran and I realized I had to get Jason out of the street.

I grabbed him under the arms and pulled him over to the sidewalk. By that time he was beginning to wake up. Lucky for us the cops hadn't arrived yet. I hurried to get Jason in the car and we quickly headed for my apartment.

The next day I told Jason he really got us into a mess. I told him if he wanted to, we would go even the score. This time we would be sober or we would take equalizers.

To my surprise he said, "Thanks, but no thanks; I'm going home to my wife and kids.

"Fine," I said, "but I better not ever see those cowards again." In my book a man isn't much of a man when he is part of a bunch that gang up on on you, and someone should educate them, even if had to do it myself.

Dick, another good friend of mine, who was also living in the apartment with us, stepped forward and offered his help. I told him, "Thanks, but it was Jason and I who got into the mess so we should be the ones to take care of it.

About a month later, I went into a bar and the bartender called me aside and told me three of the gang were down the street at a gas station.

I asked him how he knew them since he wasn't at the cafe the night the fight took place. His reply was, "I just know it's them."

I told him to save my beer and I walked out the door and got into my car and drove to the gas station. I pulled up along the side so they couldn't see me. After looking them over, I realized it was them, and then everything flashed back how they had kicked me so bad my heart felt it was going to stop every time I took a breath.

Then my cogs began turning. Since I was a customer at the station, I figured the guy in charge wouldn't turn me in. I walked straight up to the same son of a bitch who had flapped his jaws that night at the cafe.

I stayed just out of reach of him and the other two guys who were with him. They looked a little surprised to see me.

I stood there for a few minutes and then I asked him where the rest of his gang was. Like a dumb shit, he said they weren't there, and kind of laughed.

Knowing there were only three this time, I thought, "I'm going to it fast and get it over with."

I took a quick step forward and slapped that coward across the face just as hard as I could and told him, "That's for what we do to cowards and women who put themselves in a man's place."

About that time the owner of the station yelled at us to take it outside. I turned to the coward and said, "Get your ass outside." We made it to where I had parked the car, and the fight was on.

He came at me in a run and I tripped him so he went face down into the gravel. Grabbing him by the hair I smashed his face in the gravel, then kicked him in the ribs until he was begging me not to hurt him anymore.

When I figured he had learned his lesson, I asked him if he was ever going to be a chicken shit again and jump one or two people with six or more guys.

Then I told him he had better not be starting anything because he might run into a bad-ass that would kick his lights out and not just try to educate him. I noticed there was quite a crowd gathering.

Then I turned just in time for the second guy to jump on my back, but he wasn't very tough.

I dumped him and just got to thumping him a little bit when someone yelled, "Get out of here, Bud; the cops are coming."

I dropped the guy and took off out back of the station until the cops were gone.

I wondered where the crowd had come from. It had to be the bartender. He was the only one who knew there was going to be a fight. I went back to the bar and asked Ned, the bartender, how he got all those guys down at the station so fast.

He said he only made a couple of phone calls and it was a chain reaction. Each person called a friend and they all met there at the station. Before it was over there must've been fifteen to twenty people there. I felt a lot better; now those cowards would think twice before they ganged up on anyone again.

That day and night ended up in another big drunk. When I woke up in the morning I decided to leave the area and maybe quit drinking.

During the previous Timber Days, a friend of mine from San Jose asked me to come down there and live, so I decided to head south and give it a try. I got hold of Sid that day and told him I was going to take him up on his offer.

It took me approximately a week to get things taken care of and then I was on my way.

I stayed with Sid and his folks for awhile but soon we found jobs at a local plywood plant and moved into our own apartment.

I was driving hyster for the spreader in the plant and the first few weeks went by real fast. Months went by and all of a sudden Sid decided to move back home. About the same time two good looking women moved in next door.

The women came over to ask me if having a party at night would bother me. I said, "Go ahead. I can sleep through about anything." Besides, I worked swing shift and wasn't home most of the night. One of the women, whose name was Gail, said most of the parties would be on weekends.

The weekend came, and sure enough; they were throwing a *real* party. It hadn't been going on long when Gail knocked on the door and invited me over. I always kept a pony keg in the refrigerator and had already been drinking when she knocked.

I asked her if I could bring my keg along. She said, "Sure, but you don't need to bring any booze, there is every kind imaginable at my place."

I told her to give me a few minutes and I would be over.

I cleaned up real quick, grabbed my beer and stepped out on the balcony. I stood there a few minutes, just looking at the stars and drinking my beer. "What a clear sky," I thought. I looked down at the pool with the lights on under the blue, clear, still water. What a perfect night for a party.

About that time Gail and a couple of her friends looking through their window saw me standing on the balcony. They motioned for me to come over. The door was already opened, so I followed a couple of girls in.

Gail came over and started introducing me to all the different people, explaining I was her neighbor. I could only hear about half the names because of the talking, loud noise, and dancing. In a little bit I was boogieing right along with the rest of the crowd.

I had spotted a girl that liked to dance. Her name was Sharon. Sharon and I hit it off real good. We both enjoyed dancing, so we spent the next few weeks going places together.

Again, time seemed to fly; but I guess it always does when you're having fun and living it up.

It was about time for the New Year. My friend Sid, and a soldier friend from Fort Ord, asked me if I'd like to go to San Francisco for the big New Year celebration.

We picked up a quart for the after-hours bars and decided to take a hundred dollars each and go bring in the New Year.

Ken came up from Fort Ord to pick up Sid and me in his new 1942 Olds convertible. We arrived at North Beach and got started in the early afternoon, so we wouldn't miss any of the celebration.

The streets and bars were already packed with people. Before long we weren't feeling any pain and the warm

San Francisco night had closed in on the bay area. After several hours of drinking I could hardly touch my ass with both hands.

Finally midnight arrived and the people were yelling, "Happy New Year." Some were throwing streamers of paper along with papier-mache out of the high-rise buildings along the streets. Everyone was celebrating old acquaintances and the New Year.

After closing down the regular bars, Sid and I decided to hit the after-hours bars with the quart we had brought along.

During all the celebrating we discovered we had lost our buddy, Ken, in the crowd. We looked around for him for awhile and then decided we should try to find a way downtown.

We jumped a cable car that seemed to be going down into the old part of town. As we got on the car we started yelling, "Happy New Year," and before long everyone was in the New Year mood. Sid and I started singing *Oh, When the Saints Go Marching In*. Soon everyone was singing along. As we approached the bottom of the hill, we once again yelled, "Happy New Year," and jumped off the cable car.

We quickly flagged down a cab and asked him to take us to an after-hours bar. Within a few minutes he dropped us off at a real swinging bar. This place was crowded. People were dancing and still drinking to the New Year.

Sid was getting pretty drunk and he decided to go around to some of the dancers and tell them he was a Hollywood scout and might be interested in signing them to a dance act in a movie. When I realized what he was up to, I decided we should get out of there. People were beginning to get a little suspicious of him. They knew he was pulling their leg.

I coaxed Sid outside, suggesting we go find another place. We flagged down a cab and asked the driver to find us another bar close by. He soon stopped the cab; we paid the driver and headed for the front door.

A black couple was dancing as we opened the door. I made a twisting motion and the woman came over and slapped me. That angered me instantly and I told her if she wasn't black I'd probably slap her back. With that said and done we walked in and ordered a drink.

Pretty soon the girl's dancing partner came over and told us we had better leave. At the same instant he opened his sport jacket and flashed a pearl-handled razor at me.

Being super drunk , and now pissed off, I told him he better back off because the scars on my face were put there by a guy just like him; they had to pack him off after I got through with him. He must have been a real coward; he fell for the bluff.

I decided to sit in the corner booth; I thought I could at least see if he or anyone else came after me. I had no plans for leaving for anyone. Being raised in a small logging town where those things didn't happen, I realized for the first time there still were a few people out there who didn't want you in their territory.

I sat and drank for hours until I was so drunk I could hardly see. I was about to tell Sid we had better leave when a big black man, about seven feet tall walked in. A small black man was talking to him and then the big fellow came over to my booth and challenged me to go outside.

I told him, "I can't dance and it's too wet to plow so we might as well." I was too drunk or I would have taken him out right there but, instead, I managed to get to my feet and stagger outside to fight him.

I remember seeing blurs, but when it came to fighting I had been around the block a few times.

I kept dumping him to the ground but I was too drunk to hold onto him and he was so damned scared and sober, compared to me, that he would always get away and back to his feet, kicking me in the face.

Knowing how to handle myself, I kept taking him down and scaring the hell out of him. About the fifth time around I finally dumped him good.

I knew I could put the whip on him if we fought long enough for me to half-ass sober up.

About that time I started to wonder where Sid was. I couldn't see him anywhere

Then I said, "You know there are niggers and colored, and you are all a bunch of fuckin' niggers! You bastards have the nerve to call whites prejudice!"

Since the crowd was all black they didn't appreciate that at all and started for me.

About that time three or four cop cars pulled up next to the bar. After checking things out, the cops hauled Sid, who hadn't left after all, and me in one car and the big black guy in another.

They took us to the station and the officer in charge asked us what the problem was. I explained that Sid and I went into this all-black bar and a guy told us to leave, but I told him we would leave when we got good and ready. That was when the big one flashed his pearly white straight razor. I explained that was what really pissed me off.

I told him I was raised in a small town where there were no problems about where you went or who you drank with.

The officer asked me if I wanted to press charges. I

said I wouldn't waste my time; the no-good wasn't worth it.

Then I told him to put me and the big guy in a cell together and give me a knife, or take his, and I'd educate the son of a bitch.

The officer warned me to watch my mouth or he would have to put me in a cell. He asked me again if the big guy had a knife and I replied, "Did you ever see a chicken shit that didn't?"

The officer turned to Sid and asked him to try to calm me down. Sid told me to shut my mouth and let him do the talking or we were going to get locked up.

Sid and the big black guy went up close to the desk to talk to the officer. Soon they came back to where I was sitting. The officer had decided he would assign two cops to take Sid and me over to our hotel room.

We crashed. It was nearly checkout time before we got up. We were still drunk and had one hell of a hang-over. Ken had shown up at the hotel the night before.

We flipped a coin to see which one of us would go to the lobby for an Alka Seltzer and aspirin. Ken lost. By the time he came back my head felt like it was going to bust, so I wasted no time grabbing the first Alka Seltzer, put it in a glass of water, and sat there waiting for it to fizz.

It didn't start fizzing. I shoved it down with my finger. It still didn't fizz. I turned to Sid and Ken and they started laughing. Finally Sid told me that I had put the foam pad off the top of the bottle, not the Alka Seltzer tablet, into the water.

We checked out of the hotel and headed home. On the way we had a good discussion about how things were different in California cities than they were in small towns in Oregon.

I told Sid and Ken, "I don't care what color a man is — white, black, yellow, red, et cetera — he wasn't going to tell me where I could go or what to do. This is a free country and I will do as I damn well please." They said they felt the same way.

I told Sid, "You know that man I fought was the one with the good-looking black woman I ran into on the street right at the change of the New Year. I told her she had to be the most beautiful woman I had ever seen and she planted a big kiss on me."

Sid said maybe that's why he wanted to fight me. We laughed and I said, "Yeah, maybe. I wished he would of caught me half-ass sober and I would have educated him."

When we reached my apartment Sid and Ken dropped me off and went on. I headed straight for the couch and kicked off my shoes. It seemed like I just laid down and Sharon walked in and woke me up with a light kiss on the lips.

She asked where I'd been and I started filling her in on all the details. Then I looked outside the big picture window and saw it was already getting dark. I got up slowly and told Sharon I was going to clean up and then we would go do something or just stay home.

After showering, shaving and brushing my teeth, I felt like a new man. I went directly over to the couch, pulled Sharon up and gave her a big hug and kiss.

We sat down on the couch with her partly on my lap, still kissing her. After a lengthy kiss we stopped to look out the window and waved to the girl next door. She and a friend of hers were passing by my front window.

We just sat there for awhile and looked out at the still, clear night. It was so peaceful after all the New Year celebrating. I was glad Sharon was content just to be with me at home.

The next day I was back in the same old routine of going to work mixing glue for the spreader in the plywood plant. The days and weeks flew by.

9

Vietnam Beckons

One day a man came by and asked if I was George F. Brannon. I told him I sure was. He said he was an investigator and wanted to know why I was trying to get out of the draft. I told him I wasn't dodging the draft or anything else; I just moved around a lot.

He told me to report to Oakland, California, Induction Center for active duty. He handed me the orders.

The date to report was March 30, 1966. That gave me a little over a week before reporting. I had just enough time to give a week's notice at the mill. I told the foreman I had just been drafted into the United States Army.

Since it all happened so quickly, I would actually be working only three days. That would leave me time to get things organized so I could be out of the apartment by the end of the rent period.

The last day at the plant arrived and I told my fellow workers good-bye.

I went home for my last weekend in the apartment. That night I thought Sharon would show up, but she didn't. I decided to walk up the street to see a couple of girls I knew in another apartment—Barbara and Julia. Since the place where I lived was solid apartments for about six blocks, the weekends were almost like a college campus. It reminded me of living on campus in Eugene, Oregon. There were parties every weekend.

Barbara and Julia had asked me over for a private party with them if I didn't have anything better to do. I guess they weren't kidding; they were both at home.

The night went by fast; the next thing I knew Barbara was shaking me and showing me the clock. I took one look at it and jumped up and put my pants on. I kissed Barbara and thanked her for the send-off. I told her to tell Julia good-bye.

There was just barely time to make it to my apartment and grab my things and run. Sid was meeting me and giving me a ride to the bus station. We reached the station and with just enough time left to check in and tell Sid good-bye.

Since I wasn't far from the induction center, it wasn't long before I joined others and we were going through our physical checkups and testing.

Then we were lined up and every fifth guy went to the Marines and the rest were headed for the Army. I happened to be fourth, so I went to the Army.

I was put with a group going to Fort Ord, California. Soon we were loading up the bus. I sat with a young man named Pat—so young his parents had to sign special papers before the Army would take him. Pat told me he was from Oakland. He'd never been away from home much, so he stuck pretty close to me, going everywhere I went.

We soon arrived at Fort Ord and a sergeant unloaded us at the reception station. The sergeant started yelling orders for us to get out of the bus and line up.

He left no doubt he was in a hurry. He told us we

better shape up because now Uncle Sam owned us and from now on we would move faster and stand taller because we weren't on the block anymore. He informed us we would be getting a haircut, army fatigues, shots, and more tests. The testing was to find out what the Army wanted us to do.

After some more instructions about what we would be expected us to do, the sergeant split us up into several groups and put us in temporary barracks. Then we went out to get some of the routine the sergeant had promised us. When we went back to the barracks that night, I laid down on my bunk thinking, "I'm sure glad that day is in." I folded hands up behind my head and began thinking about Sharon, how nice it would be to be with her than where I was.

All of sudden, a guy started dancing next to my bunk. The bunks were close together and I was on the bottom bunk so he was shaking his ass in my face.

I asked the guy, who happened to be black, if he wouldn't mind dancing in the aisle between the next bunks over. I no sooner got the request out of my mouth when he wanted to fight.

He was yelling, "You're prejudice, man!"

He hardly got that "prejudice" bullshit out of his mouth when I came up off my bunk and said, "Now I'm telling you; get the hell away from my bunk."

I told him, "I'm not prejudice, but I'll run a head on you that you won't forget."

He started jumping around like he thought he was Muhammad Ali.

I said, "Come and get it, if you think you're man enough." I could he see he was all mouth because he stayed his distance.

So once again I said, "Get close enough if you think you can kick or hit me; I'll educate you." He just stood there so I told him to stay away from me and my bunk.

Enough of that. Soon all of us were headed for the "hill." The hill was where the training started . The first thing they did was to line us up and tell us what a sorry mess we were. Then the drill sergeant again promised we would be standing tall.

But he said some wouldn't make it because they were "candy asses." They might even be in boot camp for their whole Army career, he added.

The next day we started with early morning call, reveille. Early morning call was a sergeant walking through the barracks yelling, "Drop your hose and on your toes! Move it! Move it! Move it!"

Everyone jumped up, got into their fatigues and boots. It reminded me of early morning in logging camps. In the camps, the buzzer rang once for get-up and a half an hour later, it rang again for chow.

In the barracks part of first call was assigning details to get the work done and get men ready for training.

Groups of men were chosen to clean the sleeping area, hall, stairs and the latrine. These were broken down into more specific groups—the sleeping bay had details for sweeping, mopping, waxing, buffing and dusting windows; the latrine detail had groups for cleaning sinks, windows, floors, showers, and toilets.

It was organized so some men went to chow while others worked. When those working finished they went to chow. The ones at chow came back and did their work.

The Army kept us so busy training, pulling watch, and doing details, along with keeping our personal gear in shape, that time flew by

In about the sixth week of boot camp training, I was late getting my personal things ready for inspection. I skipped chow and worked on my foot locker, boots, brass, and the went to the latrine to shave for training reveille inspection.

I had just put on shaving cream, picked up the razor and started to take swipe under my chin when the door opened. Acting Sergeant Rusco and an acting corporal were standing there.

The sergeant asked the corporal who this guy was that was shaving. I responded, "Oh, shit" without really thinking.

"What did you say, Brannon?" the sergeant demanded to know.

I realized then that I should have kept my mouth shut so my response was, "Oh, nothing."

Sergeant Rusco told me he wanted to see me in his office when we through training in the woods that day.

When he said that it pissed me off and I told him he wasn't man enough to do it right now. I guess he thought he was. He had been an Army boxing champion of some kind.

I still had the shaving cream on my face, but I started to follow him to his room. Walking down the hall, I kept thinking, "He isn't allowed to hit me." On the other hand, I knew some sergeant had slapped Pat around. (Pat was the young man who followed me everywhere.)

When we got to the sergeant's room, I stood near the end of his bed, waiting to see what was going to happen. I was still thinking that he wasn't allowed to hit me. I just stood there, keeping my eyes on him.

He told the corporal to lock the door and leave. The corporal released the night latch but before he could leave Sergeant Rusco took off his jacket and dropped it on the bed and came at me swinging.

I stepped forward and blocked his right with my left. At the same time I dropped him with my right.

When he hit the floor I decided to do it street style. I kicked him in the ribs and took the fight right out of him.

Then I picked him up by his short collar and slapped him against the wall, using my hip to support him.

I was holding him with his head jacked up against the wall, hitting him with my forearm. I must have hit him ten to twelve times. When I let him go he fell to the floor.

I could see he was dazed, but still conscious. I told him, "You son of a bitch, you better not ever put your hands on me again or I'll really mess you up."

I pointed a finger at the corporal and said, "You saw it. He swung at me first."

He said, "Yes."

"You better not ever forget it either," I added as I unlocked the door and left.

As I began walking down the hall I met four of the regular sergeants, two on each side of the hall. The sergeant across the hall asked me if I thought I was tough.

I kept right on moving, shuffling to my left with my back against the wall, and said, "No, I don't think I'm tough but if anyone hits me they might get me, but somebody's going to get hurt besides me."

The two sergeants on my left were blocking my way but they moved and let me by.

Later, that night after training, I was called before Captain Swinger. He asked me what the problem was.

I told him I had no problem but that the acting sergeant did because he swung at me first. "Sergeant

Rusco was like a lot of people; he thought he could whip me, so he tried to make an example out of me. But it backfired on him."

As punishment the captain ordered me to pull kitchen duty every night after training. I said, "Yes, Sir, " saluted, did an about face and left the room.

June the fifth we graduated from boot camp. That day another acting sergeant, Sergeant Pepe, who was a boxer from San Francisco, asked me if I wanted to ride to Frisco with him and his family. I said, "Sure, I can catch a plane from there to home."

The sergeant was a good friend of acting Sergeant Rusco. They had put on an exhibition boxing match. With the excitement of getting leave before airborne advanced training, I never gave the fact they were friends a thought. Later I learned going to Frisco with Pepe was a setup.

Arriving at San Francisco, I got my plane ticket home. Pepe said to get it for the next day and we would party. That's what we did all right.

We went bar hopping to pick up on some women. Since I hadn't drank in some time, I got stupid drunk and never did land a woman. I got so damned drunk I was seeing blurs.

Pepe and I got to arguing about something while riding in a cab. His brothers and I were sitting in back and he was in the front.

We arrived at the bar and got out of the cab. After paying the driver, Pepe sucker-punched me.

I told him, "If you can't hit any harder than that, you better give your soul to God because your ass is mine, and I'm going to stomp the shit out of you."

His brothers spoke up. "You'll have to whip all of us."

Thinking a moment, I realized I was too drunk, so I just called him every dirty name I could think of. I told him I was out of there; I didn't want anything to do with his kind.

I went into the bar for awhile and drank some more. Finally I caught a cab and headed for the airport. It didn't seem long before catching my plane. I slept until about an hour before boarding, then shaved and washed up in the rest room.

I called Tina, an old girlfriend, and asked her to meet me when I landed in Medford. Then I headed down the corridor toward my boarding gate.

When I reached the loading gates, I cleared the metal detector, grabbed my duffle bag and headed for the waiting area until the boarding call came.

The plane was soon headed down the runway and into the air. It seemed like we hardly reached leveling off altitude when the plane started its descent. On the intercom, the pilot announced we would be landing in ten minutes.

We fastened our seat belts and the plane started its landing approach. We taxied up to the terminal in the small airport and the ramp rolled up to let us off the plane.

I started looking for Tina. As soon as I walked into the airport I spotted her. As we headed out to her car, I thanked her for picking me up. As she drove us into town, I asked about some of the locals and how they were doing. She said things were about the same as always.

When we arrived in Roseburg, Tina took me to a motel. I checked in and could hardly wait to get cleaned up. Then I got on the phone and called a few of my old drinking buddies.

That night I went out on the town and ran into several old friends. The next day I made plans to go see several friends.

It seemed like the fourteen day's leave went by fast. Soon I had only five days left before reporting to my next airborne advance training post at Augusta, Georgia.

I asked Tina to run me out to the highway; I decided to thumb my way Georgia. In three and a half days I was in Augusta.

Not long after reporting on base, I found out it wasn't a good idea to report early; I was put on kitchen duty. Since I was only one day early, I figured I could live with it for that day.

It wasn't long before some of the other boot camp trainees started drifting in, until they were all there, including Pat. He said it took those flying military standby a day longer to get to the base than it did me to thumb it.

When the training started the training sergeants told us they were going to weed us out. They said this was where they would separate the men from the boys. Only the men would be able to pass the physical examination to get into actual jump training.

The sergeants said if we weren't sure about being a paratrooper, we better drop out now and save ourselves and them a bunch of work for nothing. They said that, along with the vigorous training, we would be getting up at all hours of the night; almost like a combat zone.

The next day Pat asked me if I thought he should drop out. I told him he had to make that decision himself. Pat said he wanted to get married so he thought he should drop out so he would get home sooner. He once again asked my opinion, and I once again told him it was his decision.

The next day he came over and told me he was going to a different company. I shook hands with him and told him to take care of himself. I didn't see Pat for a long time, and when I did it was in a hospital after he had been shot.

Training seemed to be going fast. The company commander even gave us a weekend pass. I couple of Mexican friends, Mario and Pedro, and I went to town and hung on a big drunk.

I told them about Pepe setting me up after boot camp, the friend of the sergeant I whipped.

Mario said, "We don't like our kind doing that so we'll get him."

"I'll catch him myself. If I don't catch him it wouldn't really matter; most women can hit harder than he can.

He hung that sucker punch on me and it didn't even hurt or leave a mark.

Mario said he still didn't like it; "It was chicken shit."

We hit several bars in Augusta and in Colombus, Ohio, just across the bridge. We were going to a Johnny Rivers concert but never made it.

That night, after a wild drinking spree, Mario, Pedro and I went back to the barracks. We just got in before our bed check. It didn't really matter; we already had made plans to be covered in case we were late.

What could they do to us? Send us to Nam? Our orders for there were already cut anyway.

It began to seem like a lifetime before training would be over but it was only a couple of weeks. The next thing on the training schedule was an all-night run through "enemy" territory.

We went out just before dark and were turned loose in the swamps and woods. We were given compasses and told to reach a designated point. When we got there we could go back to camp and sleep.

I was able to capture several of the "enemy" and check in with the first group that hadn't gotten "caught" or "captured." I managed to get back to camp in the first truck load; that bed sure felt good.

We had only about a week to go when, one night after training, a runner Mario had sent woke me up. Mario needed me to help him and four friends whip up on a signal corps company.

I was just getting out of bed when a military policeman came running in to check the bunks. I jumped back into bed and the runner found an empty bunk to get into. The MP just walked through and flashed each bed with his flashlight, counting heads.

When he left I told the runner to go back to his own barracks. Mario and my four buddies must have already kicked ass or the MP wouldn't have showed up.

Airborne Advanced Training was soon over and everyone passed the physical test at the end of the cycle. Now we were on our way to jump-training at Fort Benning, Georgia.

When we reached Benning we started our first week of two weeks of getting oriented, listening to speeches, and, of course, physical training and running.

In the first week they repeated the warning that if you didn't think you had the guts, you better quit now and save yourself and the Army a bunch of wasted time.

Soon we were going through the ground week which was a lot of exercise and running, along with the inspection routines. During this time, we were also jumping off low platforms, doing parachute landing falls. These were called PLFs by the jump sergeants.

The next couple of weeks we were working on jumping out of towers about thirty feet high and would slide down cables about sixty feet. Then we got a day to watch the Army Skydiving Team put on a show for us.

The final week we went out of a two-hundred-and-fifty-foot tower. When the last tower week ended we got a day off to rest for our final jump weeks.

It was a nice day out and Mario and I had just walked back to the base when a group of airborne troopers were just coming back from getting their wings.

As they ran past us they could see we were still legs (regular Army), so they spit on the ground and yelled, "Dirty legs."

I yelled back, "All you guys are is a bunch of glorified legs."

The jump sergeant, who was running alongside them, had another sergeant take over and came over to get my name. That meant he would report me for extra detail at nights.

When the company commander called me into his office, he did assign me to detail. That day I had to scrape all the wax off the floor in his office with a razor blade. After finishing that, I had to wax and buff the floor.

When that was done they had me digging holes and filling them back up. When they thought I had enough they let me hit the rack a little before training the next morning.

The next week we started jump week. The first few jumps went great; then came the equipment jump. In this jump, we had our rifles and all our infantry equipment.

That morning the wind was blowing hard so the jump sergeant had the pilots make a second pass before we got the jump command.

The red light came on, then the green. Soon the C-130 was empty. I had always jumped from the flying box car, so I was eager to jump from the turbo prop plane.

The sky was full of parachutist. The first thing I did was hit on top of another man's chute. I double-timed off of it.

Then I started slipping toward the pickup zone where the trucks were waiting for us. The wind was still blowing about twenty to thirty miles an hour.

Because I was getting close to the pickup zone, I pulled back into the wind to stop my fast slipping. I stopped slipping fast, but I started oscillating under my chute when my feet hit the ground.

My chute was still moving left, but I was oscillating right. When my feet touched I started to do a parachute land-fall to my right, but was pulled back up to my left. The result was a twisted ankle.

I rolled up my parachute and limped over to the waiting trucks. As I got into the truck, I knew I was in trouble; my ankle was hurting bad.

That night I tried everything to get it better so I wouldn't be recycled. I put my foot in an ice bucket half the night, trying to make the swelling go down.

The next morning, after inspection, we headed down to the sweat box to get our parachutes. I couldn't keep up with the double-timing because my ankle hurt so bad.

Finally the jump sergeant came along and asked me what was wrong. I gritted my teeth and quickly responded, "Oh, nothing sergeant."

He told me to get back in place. Once again I gritted my teeth, even harder, and picked up my pace. Since we were almost to the ramp we had jumped off before, I just held tough.

After going into the sweat box the sergeant didn't pay any more attention to me until I jumped off the ramp.

As I jumped, I hit with both feet together, toes pointed down, so I could feel the ground and then do my parachute landing-fall left or right. The next thing I knew I was being helped back to my bunk.

The next three weeks I hung around, waiting for the next jump class to come through. When the class reached the same jump I'd gotten hurt on, I finally was able to finish the rest of my jumps.

While waiting I'd been recycled with Geraldo, the acting corporal that was in the room with the acting sergeant I kicked the shit out of in boot camp.

Geraldo and I became good friends during those week. He lived in San Jose, California, so he and I were headed for the same place for our two week's leave. I had decided to take leave there to see old friends before we shipped out to Vietnam.

When Geraldo and I got on the jet to San Francisco we thought we would get bumped. But after every stop the airline hostess would tell us just to keep our seats. Thanks to her we made it in the regular flight time.

Liquor was served on the flight. When we reached San Francisco we were really smashed from drinking all the liquor we could hold.

Geraldo's wife-to-be, Sandra, met us at the airport and we headed for San Jose. During the trip we made plans to meet again for our trip to Presidio of San Francisco to get ready to ship out for Vietnam.

The two week's leave seemed to go extra fast. Geraldo had gotten married and his new wife called to ask if I would like to ride to Presidio with them. I said, "Sure."

When we got there, it looked like a prison yard, with

its high fence around it. After a few jokes about the place I thanked Geraldo and his wife for the ride and went inside. Soon Geraldo came in.

Before long the place was full. When the shipping sergeants figured everyone was there, they called an assembly out in the yard and started calling out names to make sure we all had arrived.

As they checked us in they assigned us our bunks. We would be there a week, being issued rifles, jungle fatigues, etc. At noon Geraldo and I decided to jump the fence that night and spend some time off the base before we would be put on a bus for Travis Air Base.

Geraldo phoned his wife and asked her to meet us outside the fence about 4:30 that afternoon. We talked to a couple of our friends who agreed to cover for us and answer when our names were called on the check list.

We went out to the yard to look for the best place to get over the fence without being seen. As we slowly walked around we spotted a secluded corner where, if we had help, we could be boosted over.

When it was time for Sandra to pick us up, two friends came along and helped us over the fence. Sandra was waiting for us. We got into the car, Sandra stepped on the gas and we were speeding off for San Jose.

Sandra looked at us and said, "You guys are nuts."

We laughed and said, "What are they going to do? Ship us to Nam?"

When we reached San Jose they dropped me off at a friend's place. After a few more days we headed back to the base.

Sandra let us off in front of the gate. We said our goodbyes and had just enough time to catch the bus to Travis.

Geraldo and I figured we would be caught and punished, but we got away with it and now we were on our way to Vietnam.

10

Learning War

At Travis we boarded a stretched out D-C 8. Sitting in back, I thought the plane looked as long as a football field. The seats were all filled with troops.

Our first stop was Anchorage, Alaska. We fueled up and were on our way to Japan for the next fueling.

In Vietnam we arrived at John Sunt Airport, just outside Long Ben Army Receiving Station. The pilot taxied to the debarking ramp and opened the plane's door. When I stepped out of the plane I couldn't believe the humid heat.

We got on an Army bus and were transported to Long Ben to await orders for assignment to an outfit. Geraldo and I stayed together until the shipping date. He was given orders for Alfa Company. My orders were to report to the First and Twenty-seventh Battalion, Twenty-fifth Division, Charlie Company. I was told I would be shipped out first thing in the morning.

Just after dark a gigantic explosion went off and the sky was lit up like daylight for miles. I remember that everyone in Long Ben thought Saigon was blown off the map. The sky was a big blaze.

A split second later there was a another huge explosion. It was deafening. About ten minutes later we got word that the Viet Cong had blown up a big ammunition bunker.

Geraldo and I thanked God we had been taken off of bunker guard that night and put on latrine detail.

The next day we joined a convoy and headed to Saigon and then north to Cu-Chi Army base. The crazy thing about the trip was that they wouldn't allow us to load our weapons. We were to only lock and load when orders came from the sergeants in charge of the convoy.

About an hour outside Saigon we started catching some sniper fire. Even then they didn't give us the order to load, even though they stopped the convoy and ordered us to get off and lay down until they said it was safe to move on.

After a little air strike by a helicopter gun ship hit the source of the sniping, we moved ahead. The snipers must have gone back into one of their many underground tunnels because they were not in sight.

Finally after going through several villages, rubber tree plantations, and rice paddy field, we reached Cu-Chi. We stopped at the base gate and then rolled into what would be my base camp for about six months. The convoy rolled on inside and stopped and everyone got off at the main headquarters.

From there we were sent to the company that had been assigned us at the Long Ben reception station. This is where Geraldo and I split company. He headed for Alpha Company.

Reaching Charlie Company area, I soon learned that this was an outfit where men saw a lot of action.

Even as I checked in they were out fighting five

thousand Viet Cong. They were in the region of the Black Virgin when they approached the Cong's base camp. Eleven other companies, including air power, were called out to help in the fight.

The week-long fight was called The Battle of Attleboro.

At the end of the week I had already pulled my first ambush, just outside of Cu-Chi Base Camp's wire and mine fields. We only went out into the rubber plantation to lie in wait for the enemy, but it was for real.

Some of the sergeants decided I would make a good point man so that became my first job. They thought I did so well that they kept me as scout the rest of the time in Vietnam.

The first week went fast. The company was alive with new men who had come to replace the dead and the wounded.

One of the new men asked me how it was running point. I told him I would rather do it than have someone filling the job who may lead our squad, platoon, or company into a Viet Cong trap.

He introduced himself as Ray and said he would be the scout out front, running point, that night and just wanted some advice.

I told him, "Ray, the first thing you want to do is get off that funny grass, be alert." I could tell he was smoking marijuana.

"Now, they are going to tell you not to put a round in your gun until you see the enemy or are shot at, but you are going to be better off if, when you walk through the wire and mine field around base camp, you pull back the bolt of your rifle and slide a round into the chamber.

"Then put the safety on and keep your thumb on it so you can flip it at anytime and be shooting fully automatic. That way you will always have your trigger finger on the trigger."

I added a warning. "You better pay attention to everything I told you."

I asked him if he remembered seeing a skeleton the Cong had chained to a tree. They had captured the guy and put a steel collar around his neck and chained him to a tree. When the shooting grew more intense they shot their captive and ran to their tunnels to hide.

The guy was probably just a South Vietnamese rice grower. We figured they killed him to intimidate others like him in the area.

Ray was really scared; I guess he had probably not been away from home much. I couldn't help being sorry for him, he was so damned scared.

I told some soldiers Ray shouldn't be running point, especially since he was smoking funny grass. It looks like I was right.

Later that night, the platoon he was assigned to went outside the wire and mine field for their first ambush. They had hardly got out and walked up on the enemy when Ray was shot in the head. When I first heard about it I was pissed off because I knew he shouldn't have been running point.

I didn't have much time to think about it because we got orders to get our jungle gear packs together. About ten of us were on our way to the jungle as replacements.

We boarded two Huey helicopters and started on our way. When we got off the chopper a sergeant was waiting to tell us what to do next. He pointed out the platoon lieutenant and said we were to get our placements from him. Reporting to the lieutenant, we were assigned to the platoon we would be in for the rest of our Vietnam tour of duty.

That afternoon we headed for the jungle on a search and destroy mission. We only encountered some occasional snipers.

In the evening our platoon lieutenant gave Pat, who was a buck sergeant, and me orders to be on observation point on high ground away from the rest of the company, to watch for any approaching enemy.

We got a fox hole dug just before dark. Then we encountered sniper fire.

The two men sent with us to help pull guard jumped in the hole on the left and got down as low as they could, covering their heads with their hands.

I was standing on the right side of the hole, looking for snipers, when I glanced at Pat. He just gave me a wave of his hand. I wondered about the reaction of the two guys who put their hands over their heads and jumped in the fox hole.

After the shooting stopped, Pat told me I would see some of that over here. "Some fight and some don't, of all nationalities; I guess it's the way they are raised," he explained.

I later found out he was right. A lot of men just froze they were so scared. Some just laid flat and forgot to shoot back.

The next morning we got orders for our platoon to head north on Eagle Flights. The idea was that a helicopter would drop you off to clear an area of the enemy and then move you on to another to do the same thing. They put us down in an area where spotter planes had seen Charlies moving. *(Charlie* was a name we used for Viet Cong soldiers.) It was thought that these Charlies were retreating from Attleboro.

That evening I asked Pat how they clean up here in

the jungle. Pat told me, "When the rain hits use your helmet to get as much water as you can."

We had just dug in when the rain hit. I stripped down and started taking a shower in the downpour.

Then Pat started shooting. I figured I was going to get a chance to use the new M-16 a soldier had just given me because he was going home after a twelve-month tour.

I grabbed the rifle and jumped into a foxhole, totally alert. The foxhole was filled about waist high with water. When I jumped in, water and mud splattered all over me.

Standing there nude, covered with mud from top to bottom, I started to look around for the snipers.

I thought I saw a couple. I started to shoot back, but my gun wouldn't shoot.

I was one pissed off soldier—my gun wouldn't work and I was covered with mud. On top of everything else, the rain had stopped and I had a hell of a time getting all the mud off me.

Standing there like a jackass in the nude, I got my canteen and tried to rinse off with what little water I had been issued. That was the last time I tried using a rain downpour as a shower; I felt dirtier afterwards than before.

After things calmed down I told one of the leaders that the guy who had just rotated must have hidden every time shooting started, because his gun had never been fired. Lucky for him he was already headed for base camp and home.

I asked a sergeant if I could get some steel wool and gun oil. He took me over to a supply sergeant and told him to give me some.

Then I started taking the clips apart and sanded the

springs. A couple hours and several sore fingers later, I got the springs to shove the bullets back up into the chamber.

During the rest of the week I went on several ambushes, but never ran into Charlies. The company commander decided they must have moved out after the main Battle of Attleboro, so he took us in for resupplies and more replacements.

I thought we would get a break when reaching base camp but it turned out to be work as usual. The people working in the camp got the break while we took over their jobs. Also we pulled ambushes outside, then came back in and filled sandbags all day.

On my fifth ambush I was cleaning my gun when a man walked up to me and said, "Hey, Brannon." I looked up but not recognizing the man I went back to work. He again said, "Hey, Brannon," but I still did not recognize him.

I was too wrapped up in getting ready to go out on another ambush, and I was also too damned tired to care.

The third time he called me I looked up. This time I recognized him.

"Well, I'll be damned if it isn't old Don from home!" I almost yelled.

He told me I better get out of this company because it gets more shit than any other company around. I only got to talk to Don a little. I had to go to a meeting, but it was sure good to see somebody from home.

About that time I heard one of the cooks walking by. He had a radio. It was playing a song called "My Baby Wrote Me a Letter." My girl in San Jose came to my mind. Then my train of thought was broken by the sergeant yelling for everyone to form up.

I had never seen the new sergeant who called the formation, but it didn't take long to find out he was a by-the-book tusky. Back home that's what we called people who didn't have much common sense.

The sergeant started out telling us we were going to be pulling two-man positions on the ambushes. Pat, who had been in Vietnam almost the whole tour of duty, took issue with that. He told the sergeant we were going to pull three- or four-man positions.

The sergeant bristled and told Pat that he was giving the orders.

Pat tried to explain to the sergeant that this was our fifth ambush and that when we came back in, we were either made to pull line or taking care of other duties, and we need the three- to four-man positions so we could take turns getting some rest.

The E-6 sergeant got pissed off and ordered Pat to report to the company commander for a court martial for not obeying orders.

"If we were in the States we would have ways of taking care of guys like you," Pat said.

He asked Pat if that was a threat, and Pat told the sergeant, "No, that's a promise."

Gill, Dave, Lupe, and I joined Pat and we were all headed for the company commander's hut when our lieutenant came out with the ambush information.

He asked us why we weren't back with the others. Pat told the lieutenant we weren't going to pull two-man positions.

We started walking to the formation with him and he gave Lupe and me the coordinates and distance for going out on ambush.

Then he called the new sergeant over to the side and

told him to take our advice, that we knew what we were doing, and that this wasn't training it was for real. After that the new sergeant never cared for the five of us.

That night we went out the wire through the mine field where the observers spotted some Charlie activity. During the week we had several mortar attacks on base camp and it was believed they came from where we were going to patrol and set up our ambush site.

It was especially dark; I remember having to stop and light a match under my poncho just to get my compass to illuminate so I could see the degree marks.

After I got out to the ambush site I waited for my backup compass man, Lupe, to come up so I could double-check my reading.

I remember thinking how dark it was. It reminded me of one night back home, hunting in the mountains. We had a hard time getting home because it was so dark, like tonight. I couldn't see my hand in front of my face.

Because it was so dark in the jungle, all kinds of things were going through my head: "What if I walked straight into Charlie on the trail?"

Back in my mind I knew I could kill him, even with my feet and hands if I had to. But I thought I would probably react by opening up with my M-16 on fully automatic.

My train of thought was broken up by Lupe arriving with word that Lieutenant Gravel said for everyone to lie low. This is where we would find out if I was out the right distance by my compass and count.

The first was a marking round fired by the company artillery and the second was a high explosive round, a phosphorus flare round that helped fake out the Cong and find correct distances. They both hit fifteen meters in front of us so I must have been real close to the exact distance from base camp.

We set up our men for the ambush and waited for Charlie to come along. Because it took us most of the night to get out this half-mile or so, we didn't have but one or two guard sessions to pull before heading back fairly early.

On our way back it was easy to see anybody or movement of any kind. After several hours of traveling, we were getting fairly close to base camp.

Lupe came running up and told me to swing to the left. I asked him why. He said the lieutenant figured I was going to miss the gate back through the mine field and he didn't want to hit any mines or flares because our own line could think it was Charlie and open up on us.

I told Lupe that the lieutenant must be off a little in his direction because I was going by my compass, and I wasn't changing.

I asked Lupe who he was going to follow and he said he was going to follow me. I told him to go back and tell Lieutenant Gravel that I was going to stay going the way I had been and that he should follow me. But the lieutenant ordered me to change direction.

I asked Lupe again who he was going to go with and he said with me.

I went about twenty meters and could see the wire and the path going into camp. I motioned to Lupe to show him I was right; I had found our way back to camp directly on the nose.

Then I told him to go back and catch the lieutenant before he led the rest of the platoon into a mine field. Lupe told the lieutenant I had hit the gate right on.

From that time on I was never asked to change my

coordinates. It reminded me of a time when I was a boy hunting in eastern Oregon. I had lost my direction but quickly found my way back home.

Back in camp it seemed we always had something to do. Another new E-6 sergeant came up to me and said I would be pulling line that night. I told him I couldn't because I hadn't had any rest for a week because of all the ambushes we had been out on.

He said I better do it or he would have me put in the stockade. Then, to make it worse, he put me in the main bunker. In the main bunker you had to call all the other bunkers in your company area every half an hour to check on their activities. After each check with the bunkers you had to report what you had found out to battalion headquarters

Well, I fell asleep and when I didn't check in the sergeant was sent down to get me. When he found me asleep he was pissed.

I said I had told him I couldn't stay awake.

The smart bastard grabbed for me to take me to the company commander. That really pissed me off. I warned him to keep his hands to himself.

Lucky for him, and also for me, our platoon lieutenant came along as we were going to the company commander's hut. The lieutenant told the sergeant he would take care of the problem.

After telling Lieutenant Gravel I had warned the sergeant I couldn't stay awake, the lieutenant said for me not to worry about it. He sent me to my hut to get some sleep.

That experience made me realize there are real men in every position, but there are also some who don't deserve to be called men.

I guess the lieutenant didn't much care for the sergeant; I never again saw the sergeant in our company. In an infantry company there were always other places for men who thought they knew it all. This was the second sergeant we had that tried to go strictly by the book.

The men in charge knew they had to have the complete cooperation of the men under them because the fighting men are the backbone of the armed forces. We were damned lucky to have some sergeants and platoon officers who understood they had to have it that way to keep the men's minds on doing their jobs.

That night I got a good sleep in a real bed. The next day Lieutenant Gravel sent a runner to tell me to take the day off. I got a haircut and decided to check out Cu-Chi and just take life easy.

In the evening I decided to take a real shower. Getting some clean jungle fatigues and my thongs, I headed for the company showers. The showers were outside; canvas skirting were the walls. It reminded me of camping back home; building an outdoor shower in the woods by hanging skirting of plastic around four trees.

Then in a flash my train of thought was broken by somebody yelling, "Incoming!"

The next thing I heard was explosions around Cu-Chi. At first I thought of running for cover. Then I decided to keep on enjoying my shower because it wasn't often I got a shower with all the water I wanted.

At that moment one of the other men in a shower stall yelled, "Aren't you coming?"

I said, "No, piss on Charlie."

After a few more explosions it was all over, and I got a nice long shower.

Leaving the shower I couldn't help but wonder if

anyone was hit. I guess everyone made it, because I didn't see any medical personnel in the area.

Thanking God for protecting me and everyone in the area, I went to my hooch, which served as our barracks. As I entered, Pat, our buck sergeant, said we were moving out in the morning.

I sat down on my bunk and started shooting the shit with Jim, a black man who was new to our platoon. Little did I know he and I would become good friends.

Soon the hooch was full of other men from our platoon, sitting around the bunk, smoking cigarettes and shooting the shit.

11

The War Gets Real

The next morning we were up and headed out the Cu-Chi gates in a convoy north. The trucks went through the little town of Cu-Chi. The youngest kids in the town ran alongside the trucks, with their hands stretched out, begging for food. They looked like skin and bones.

We had been issued C-rations for the day so most of the soldiers threw them to the kids.

Then I noticed one soldier throwing his C-ration cans, trying to hit the kids and yelling, "You no good Viet Cong."

I told him, "You son of a bitch, if you hit one more kid I'll kick your damned head in. From then on I never took my eyes off him anymore than I had to.

After a couple hours of traveling we came upon five Viet Cong stretched out alongside the road, killed by South Vietnamese soldiers. I was told they displayed them to make other North Vietnamese stop and think twice before trying another raid.

We reached our destination in about another hour. As we started getting off the trucks Lieutenant Gravel said we would be riding some choppers out to the Black Virgin area where we would be setting up base camp.

First we would pull a company sweep, walking side by side in a line through an area, looking for Charlie. The lieutenant put me running point, gave me my compass reading, and I headed out.

After walking for a couple of hours I saw a village way off in the distance; I could barely see it. We were being shot at from the village. I could hardly believe it because it was too damned far away; they couldn't possibly hit us.

Then it dawned on me; it was just to make us stop so they could get men or equipment out of sight. It worked; when we got into the village all the people were gone. Searching out the village, we didn't find anything.

We had finished the search and were getting ready to leave when we started getting small arms fire again. It was from so far away the captain, our company commander, didn't pay any attention to it. We continued our sweep.

When the company reached a road where we were to meet the choppers, we were air-lifted to a place in the jungle to set up a company base camp.

That night, as always, it seemed, our platoon was the one sent out on a night ambush. We went to the ambush site early. Then, in early morning, we headed back to base camp.

When we got close to camp one of the M-79 carriers accidently shot off a star cluster. It lit up the whole area. The platoon hit the deck.

Lucky for us our lieutenant had just called in and said we were coming back to camp, so the company perimeter guard didn't open up on us.

When the cluster finally went out the lieutenant told the perimeter guard what had happened. Then word to move out again was quickly passed up to me. I got up from where I had hit the dirt, and took the platoon into the company area.

That day we finished laying out flares and rolls of barbwire to fill in gaps around the perimeter. That night we could see the Fifth Mechanized Company, a personnel track unit, in a firefight off on the western horizon. Some "Puff the Magic Dragons," helicopter gunships that could fire over a large area, were out there also; they were bombing the piss out of the jungle.

Lieutenant Gravel came back from a headquarter's meeting and told us we were going in to push Charlie back because Charlie had the personnel carriers surrounded. Lots of the tracks had run into mines.

The place was known as Hobo Woods and because of the thick jungle there were lots of Charlies in that area. Our buck sergeant told us that when we landed we would move out in front of the tracks and then lay down and wait for the orders to push Charlie back.

The helicopters arrived and we boarded. Soon we took off for the firefight in Hobo Woods. Before long we could see the tracks.

Three of them had formed a big circle around the tracks that were demobilized. It reminded me of the movies of circled wagon trains in the Wild West being attacked by Indians.

Planes and Puff the Magic Dragon were hitting jungle about twenty to fifty meters in front of the tracks.

The fifty-caliber machine guns on the tracks were cutting down the jungle. Charlie was dug in so he was hard to run out.

We set down in the middle of the personnel. When the helicopter touched down I was on the left side, but for some unknown reason I moved over to the right side and jumped out.

Just as I jumped a big explosion hit the left side. It

wounded the pilot and killed Ben, an infantryman. It seemed like it really didn't happen; one moment Ben was there, the next moment we couldn't find any part of him.

On the ground, all of a sudden some lieutenant started yelling, "All of you men get over here!" Since he wasn't our platoon leader we didn't pay any attention to him.

Then our first shirt (the top sergeant) yelled, "Stay down until I tell you to move out!"

The next thing I knew the first lieutenant and our first shirt were yelling at each other.

The lieutenant said he gave us an order to move out with him and that's what he wanted us to do.

The first shirt told us he was our platoon sergeant and he didn't want us to move out until he got us out in front of the tracks and everyone was in position for the sweep.

The lieutenant got pissed because no one was doing what he told them to do. He told the first shirt he would have his stripes for not backing him and his orders.

That's when the first shirt got pissed and told the lieutenant, "Your new here and don't have anything to do with my men. You can shut up or I'll have your bar."

After that our first sergeant moved us out on the left side, in front of the tracks. Since I was a point man he said he was putting me on a cross-trail in a piece of jungle running out to the middle of the opening where the tracks were in a circle. The rest of the jungle was on the outside of the tracks.

We stayed down because of the firing and bombing raids.

After lying there for awhile I thought I heard something behind me so I told Rod to keep a watch on the front trail. Taking my rifle I walked back on the trail.

After going about ten meters I saw where Charlie had

been digging round holes to bury mines to knock out the tracks.

I looked around cautiously, but didn't see anything. I still thought I heard something.

I went back up to Rod and told him about the noise I had heard.

All of a sudden a fifty caliber opened up behind us. A man came running and yelling, "Get out of there; Charlie's behind you."

As we moved back, the track gunner shot and killed four Viet Cong as they came out of their tunnels. They had killed the track driver.

Then another track gunner detected where the shot came from and opened up with his fifty caliber machine gun.

When we got back to the tracks Pat, Rod, and I went in and pulled the four dead Cong into the opening. Pat said he had a French Vietnamese girl back at Cu-Chi who wanted some ears of the Cong.

She probably didn't think he could or would really get her some ears. He took out his knife and started cutting the ears off one of the Cong.

Then he went back to the tunnel and blew the hole with an explosive called C-4 you could form with your hands, and we started moving back to our positions on the jungle trail.

We waited there for a considerable time; dusk came and the Cong seemed to have moved out. The company commander called off the sweep. Choppers came in and flew us back to our base came in the jungle.

The lieutenant assigned us our guard duty for the night. I was real lucky; I only got one hour of guard.

The next morning, Pat said we were moving out. He

also told us the sergeant who said we were going to pull one- and two-man wouldn't be with us anymore.

I asked why. He said, "The dumb shit went out front of one of the line positions last night and got shot."

I asked, "Who shot him?"

"I took over the machine gunner's forty-four and shot him in the legs," Pat answered simply.

"I was surprised. "Won't you get in some kind of trouble?"

"No, the machine gunner and I yelled for him to identify himself. He didn't answer so I shot him with the forty-four; he started yelling, but it was too late.

"I felt bad, but at least he's still alive. And he won't be able to get a bunch of our guys killed with his dumb decisions," Pat added.

After breaking base camp, we loaded on choppers and were flown to the Delta. On the outskirts of the delta country we began setting up another base camp.

We had been digging in for several hours when I heard a chopper fly in. I was about to finish sand-bagging when a general came by my position.

I stood there for a moment, caught off guard. Then I realized it was the big man himself so I quickly saluted him. He stopped and said, "You men are doing one fine job," at the same time saluting back. He told us to carry on and moved down the rest of the positions.

I finished the fox hole, sat down, and pulled a C-ration cigarette out of my elastic band holding my camouflage cover on my steel pot. Lighting up, I sat there puffing on the cigarette and talking to Gillman and some of the other soldiers. Night came and we pulled our usual guard duty.

The next morning our lieutenant came by and told us

we would be going on a sweep in a few minutes. Soon a runner came and told us to form over by the lieutenant's tent.

Lieutenant Gravel gave me the compass information and told me how far out to go. In just a few minutes we were moving out on the sweep.

The going was easy at first, then I ran into elephant grass and waist-deep water. The grass was so thick I would lunge with my rifle held at chest level, cross ways, trying to knock down the grass to take each step. This went on for several hours.

That afternoon, Pat, our buck sergeant, came up and asked if I wanted a break. I told him I was fine and would make it all right. Pat said he was just checking because men behind me were passing out and helicopters were being called in to pick them up.

I kept pushing and hoping to come upon a dike for a slight break. About the time I had accepted that there were no dikes in the Hell Hole, I finally came to flat ground. This was one time, out of many more to come, it took lots of backbone to make it.

Going on I could see a village in the distance. When we got closer enemy snipers started shooting. We returned some light arms fire.

It must have got too hot for Charlie; he moved out.

We moved into the village and lined up all the people. The captain and lieutenants checked their South Vietnamese ID cards. It sure seemed strange to me that the people were either old or real little kids.

They all had ID cards so we moved on.

We hadn't gotten far when we got shot at from behind but no hits. I thought the Cong were poor shots or they just wanted us to think the villagers were Cong.

We moved through jungle and across another opening where we reached a road where trucks took us back to our base camp in the jungle. Reaching our camp, I realized we had just made one big circle that day.

We were in the Delta area for some time. This was normally Big Red One's region. Most of the time we used Eagle Flights to search the north section. We would be flown out to an area and sweep it, then we would call in choppers that would fly us to another area, to check it out.

We weren't getting any heavy contact so we were sent back to Cu-Chi to resupply.

When we got back, our company was picked to pull line duty. I was put in the main bunker. I had to call all the bunkers on line in battalion sector and then, every hour, call battalion and give a report of the situation.

That afternoon I called Jim's bunker on the routine check. Jim said that he wanted me to come down as soon as I was through making all the half-hour checks I signed out and said, "Okay."

I finished my sector of bunker checks and ran down to Jim's bunker. He had two French Vietnamese girls. They both spoke some English and were real pretty.

It dawned on me that we could get in big trouble if we were caught with them so I told Jim I had better get back to make my hourly battalion report, but Jim said he would switch bunkers with me and make my routine bunker report so Susie and I could have some time together.

Jim came over to my bunker, just as he said he would, and so I took off to watch from his bunker. Susie was waiting in the bunker, sitting on the bed, and we started talking.

I asked her if Susie was her real name, and she said

it was in the English language. It was real nice to have a girl to talk to.

After a short time of just sitting and holding hands, she leaned her head on my shoulder and started kissing me on the cheek. Before long we were embracing each other. Then she was asking me if I would make love to her.

Lucky for the two of us the bunker had a place to sleep. After we were through, Susie asked if I would be coming back. I told her I doubted it because I was infantry and we usually were out in the jungle looking for Charlie.

I glanced at my watch and realized I better get back to the main bunker for my battalion report. As I reached the bunker I thanked Jim for introducing me to the French woman. Then I asked him how he got them up there, and also where he got his hands on the condoms. I had to wait to hear the story because Jim had to get back to his bunker.

Calling battalion I gave my bunker report. In a half-hour it was time again for a bunker check. Calling the first bunker, Jim answered by saying, "This is Major Jim."

Without thinking I said, "This is General Brannon. Is everything clear?"

Jim reported, "Everything okay. Major. Jim clear."

I said, "General Brannon clear."

When I called the rest of the bunkers they answered in the same way, as lieutenants, captains and so forth. After making bunker checks a couple of times like that, battalion sent a runner to my bunker and warned that we'd better knock it off or Charlie would pick it up and hit base camp with a barrage of mortars, trying to get some big shots.

Not to mention the fact we could get court-martialed.

Thinking back, I was sure glad we hadn't started calling each other big brass titles before Jim and me were with those two French Vietnamese women. That runner may have caught us with them.

That night, when we got to company area, our platoon Lieutenant called us in and told us not to do that anymore or we would get in big trouble.

Charlie hadn't been hitting us with mortar fire. The brass figured it was because of a cease-fire to observe Tet, the Vietnamese Lunar New Year.

Lieutenant Gravel called a platoon meeting. He said that tomorrow Bob Hope and Ann Margaret would be in Cu-Chi, but we would be on ambush and patrol while they were in camp.

A few guys asked why camp personnel weren't sent out so we could watch the show. They thought we ought to see the show because the clerks had easy living compared to infantry men.

The lieutenant said he agreed, but the clerks wouldn't know what to do if they did see a Charlie. "Sorry," he said, "you're picked because you guys are the backbone of the Army."

That day we got our rifles, ammo, and gear ready to leave that night on ambush patrol to protect Bob and Ann's show.

The cease-fire was still in effect. But that didn't mean much to Charlie when you ran into him when he was packing a rifle and probably high on the pot (marijuana) carried in baggies tied around his waist.

I took point again but this was an easy one; we only had to go out to what would be mortar range. When I reached my destination point I changed my mind; this was not going to be entirely easy.

It started to rain a cold monsoon rain. The water came down like a bucket pouring out water that had no end. The rain kept up most of the night. I tried to find a high spot because the water laid in pools everywhere.

Finally I found a high spot. I crawled up on a mound of dirt and even it was soft. That night was a long one, trying to cover yourself with your poncho and get some sleep when you were through with guard duty.

Morning came and the rain stopped. In the early morning light I discovered it wasn't the rain that made the mound soft; it was an ant hill.

The lieutenant passed word down for me to get over to him because we would be searching the area and I needed to get the compass reading from him. When I had the information we started our night search at Mortar Camp.

About twenty hundred hours the lieutenant sent a runner up and stopped me. The runner said the lieutenant and Pat would be going out ahead of the platoon to check a report of possible Cong movement.

Pat showed up. He had killed three Cong who were sitting in a hooch with their rifles and were high on grass.

Then the lieutenant and Pat came back and told Gillman, Lupe, Jim and me to go with them. The lieutenant left and E-6 sergeant in command until we returned.

When we got to the hooch, the lieutenant said we had to bury the Cong and keep it quiet because it was supposed to be a cease-fire.

I remember thinking how I hated digging fox holes and this was a lot bigger. Five of us started digging; lucky the ground was so soft. It didn't take long.

Standing there looking at the graves, I thought, "There's three that can't get Ann and Bob with mortars. Then my train of thought was stopped short by our lieutenant yelling, "Men, now we better find the two who got away."

The five of us started searching and we were able to track some blood to a few cave entrances, but it looked like suicide to even try to go in them. We headed back to the platoon.

We spread out and combed the area about half way back to camp. The lieutenant had me take point the rest of the way.

Back at camp we found Bob and Ann had come to Cu-Chi and were already safely gone. Lieutenant Gravel said that since we didn't get in on the show we had the rest of the day off to do anything we liked.

I remember it was a real treat. I showered and got a haircut. Gillman and I decided to walk around and check out base camp. Down by the battle-line headquarters we found the Army was having a show, without Bob and Ann.

It was just some men and women in the Army, but it wasn't bad. In fact, when a bunch of girls came out, it seemed like Hollywood. It made me think of home and what might happen when I got my R & R vacation in Australia.

At the end of the show I checked out the rest of the camp and then kicked back to relax the rest of the day.

12

An Attack That Worked

A while later our sergeant came in and told us we were moving out in the morning at 6 o'clock. We had an early breakfast and were issued extra ammunition and grenades.

Soon trucks came to pick us up, Before long we were on our way to Cu-Chi airfield. We stopped at the runway and everyone got out of the trucks.

I sat there wondering where we were going. I asked others and it seemed like nobody knew. But it wasn't long before the sky was filled with helicopters coming from the south.

Lieutenant Gravel finally came to fill us in. He said we were going north to hit a village of the North Vietnamese regular army. A platoon had already been sent in and three soldiers were missing. They were figured to be prisoners of war in the village.

He issued us boot laces and said to tie everything we had, that was stamped U.S., to our belt loops. We were going into neighboring Cambodia. U.S. raids there hadn't been officially authorized, although the North Vietnamese had been using Cambodia as a base for ambushes into South Vietnam, so we tied on anything we might drop that could be used to prove we had been there.

Then the lieutenant opened a map and showed us what the village looked like from the air; it looked like an oasis in the middle of a large lake.

He wished us luck and said that hopefully we would find the soldiers quickly and maybe we wouldn't have to take as many casualties as the first attack on the village.

When we got over the village we were supposed to drop all around it and use bandoleer torpedoes to blow holes in the surrounding hedge.

The choppers landed and we were ordered to board quickly. When we were airborne it reminded me of jump training in Fort Benning. I noticed one big difference: we were flying very high until we were over the Saigon River, where we dropped down so the choppers' runners looked like they were only a few feet off the ground.

Soon I could see the "oasis" in the distance, then we were there and the choppers came to a fast stop.

We were ready to assault the village, but the lieutenant was in the chopper in front of us and no one was getting out of it.

I looked at Pat and Gillman. Charlie was shooting the shit out of the chopper. I said, "This is where we were suppose to get out; why isn't anyone getting out of the chopper."

Then Pat said, "Jump!"

The major flying the chopper yelled, "Jump, you bastards, jump!"

Before the flying major was through hollering, all five of us were already out of the chopper.

The front chopper stopped again. The lieutenant jumped out and ran over to us. Charlies were running everywhere.

We started shooting at Charlies as, in a staggered

line, we ran to the village as fast as we could. This way, if one of us stepped on a mine it wouldn't get the others.

With all the running and shooting at Charlies, I didn't give the mine fields a second thought.

Charlies were shooting at us, and the ones who weren't were running for hooches to get their weapons.

As we entered the village five Cong ran across in front of us to our right and headed for their hooch.

When they ran into the hooch, I followed them, hoping to get them before they got their rifles.

I moved slowly and checked out several rooms separated by thickly woven walls of elephant grass mats. It seemed they had vanished, but I knew they must have gone in a tunnel to get their weapons.

I lifted mats with rice and pots on top, anything that might have a trapdoor underneath. All the time I stayed off to the side with my M-16 on fully automatic.

After a quick check of all the rooms I backed out, facing the way I came in.

I knew Lupe was covering my back. Sure enough, when I came out, there he was, standing off to my left. I looked at him and shrugged my shoulders, as if to say, "I don't know where they went."

I turned my back to the door and started to go around the right end of the hooch when Lupe yelled, "Watch out!"

As I turned back towards the door, five Charlies with rifles dashed out. They ran between Lupe and the hooch and headed in my direction.

I figured they would have to run across in front of me and the rest of our men who came running from other hooches.

They got about seventy-five yards away before our

combined fire dropped the Charlies. They appeared to be dead.

I ran over to them, but Pat said, "Get back and finish them off with a twenty-round clip." He was warning me they still might be able to pull pins on their grenades, so always make sure they are dead.

It hit me then: Pat had been in Nam so long he was getting jumpy. They say when a man's Nam time gets short, some get really nervous. I figured this would probably be his last operation.

Anyway, after making sure the Charlies were dead we took their weapons and started searching for the three captured Americans.

We were in the center of the village, running in and out of hooches, looking for the Congs. We could see them running all over the place when we first came in on the choppers. Most of them were shooting at us and running for tunnels, just hidden holes.

We opened tunnel hatches and hollered for them to come out before we blew up the hole. Before long we had two chopper loads of regular Vietnam soldiers from the north. These were trained soldiers like ourselves.

After getting them altogether we told them to sit down in a squat position with their hands over their heads.

All except one did what we said. Every time we would shove him down, he got back up.

Pat got pissed and told him he better stay down, but he wouldn't.

Pat started using his rifle butt to knock around the one who wouldn't stay down. I told Pat I would make the Charlie sit. I think the Charlies were scared we were going to shoot them.

I walked over to the Charlie and motioned for him to squat down; I even did it myself while putting one hand over my head so he knew.

Then I reached in my pocket and pulled out a cigarette and gave it to him. I guess he figured everything was all right. He squatted down and smoked the cigarette.

The lieutenant told me to load them up in the helicopter while the rest of our men searched other tunnels.

At first the Charlies didn't want to get in the chopper, but I motioned with my rifle upward and they started to cooperate. The helicopter was packed full..

I talked to the pilots and told them the Charlies were scared we were going to kill them. The pilots didn't seem to care.

One of them told the two gunners in the back that if the Charlies got sick, to kick them out. I thought, "What a stupid-ass son of a bitch; I'd like to kick *you* out."

I headed back to where the other men were and ran across a woman who had been shot in the knee. It looked like the wound could be fixed. I decided to carry her back with me to see if I could get her any medical help.

About this time Lieutenant Gravel told me we had to load up in the choppers and that I couldn't take her. I told him I would carry her, but he ordered me to put her down.

I motioned to her with my hands that I had to move out. I laid her on a table in a big hooch.

We had a few minutes until we had to leave so I quickly checked out the hooch. It was like a big classroom, used for everything. There were rows of benches with a big blackboard in front. Off to one side I found an old pedal-operated Singer sewing machine.

The lieutenant called for everyone to board; we were on our way back to Cu-Chi. When we arrived the lieutenant had us wait for the choppers to leave, then we packed our stuff onto trucks to go to the company area.

This operation had been a big success. Because some of us had charged the village line and had dropped the rest of the company around it, we put the North Vietnamese regulars on the run like rats; they couldn't get in their holes fast enough.

As for the three missing Americans, the brass figured they had been taken to North Vietnam.

Because the six of us reacted and did what we were suppose to we didn't take one casualty. I was told battalion was going to use the operation for Army training books.

I would call us, "The Magnificent Six." I told the men what a good job they did. I was proud of them and glad everyone came back in one piece, and most important, alive.

The men were told to stick around till we got orders for the night. I said I would try to get them the next day off. The attack on the village was a job well done and we all deserved a rest.

13

Pinned Down

In about an hour Lieutenant Gravel came back and told us we were moving out in the morning and would be on a convoy north, but we could have the rest of the day off.

That evening, because we felt lucky to still be alive, the Magnificent Six had a few beers on base. After a couple of hours we headed back to our hooch. It was unreal; a good old Army bed again instead of the jungle. It was great to crawl in the sack early; I had hardly hit it when I was sound asleep.

Morning came too quickly. We got our gear together. Soon the trucks came, our company loaded up, and we were on our way north. Word was passed along that we were on a big operation up north called Operation Iron Triangle.

Word was that lots of Cong had been spotted moving around in the area where we were headed. This was supposedly the second day of the operation.

We rode for about three hours and then the convoy pulled off onto the side of the road. Lieutenant Gravel came along and ordered us off the trucks and said to spread out.

Soon we saw phantom jets scabbing the jungle off in the distance. Then the word came down from the lieutenant that we would be going in to help our sister outfit, Alpha Company.

The lieutenant said the outfit had encountered heavy weapon fire while sweeping the jungle. He told us to have our weapons ready when the choppers came in.

In about ten minutes I could see the helicopters. Then I could hear the chopping sound of their blades and they were landing alongside the road. Lieutenant Gravel said we would be the last platoon going in.

Troops started loading into the choppers, and as soon as they took off we could hear the lieutenant saying that we would trap the Viet Cong between us and Alpha Company.

Soon the choppers were coming back to pick up more troops, loading, and taking off again. It seemed they had hardly left when they were back again to pick us up.

The lieutenant passed it down to load as soon as the helicopters were on the ground; the pilots nodded their okay. We loaded in what seemed only seconds.

The choppers lifted off and headed for the rice paddy behind Alpha Company, part of the plan to get the Cong trapped between the two companies.

As we reached the rice paddy, the choppers made a pass because the other platoons and Charlie were engaged in a hot firing battle. It was simply too hot to go in. Some of the replacements for guys like Pat, our old buck sergeant who got an early discharge, said, "Oh, boy, we don't have to go in."

Lupe looked at me and I gave him the same look, as if to say, '*Sure*."

The choppers made another pass and came back to

hit the hot landing zone. We jumped out and started charging the jungle line.

I turned and noticed the rest of the men back at a dike behind us yelling, "Come back! Come back!"

Looking to my left I saw the new buck sergeant who had taken Pat's place. I yelled, "What the hell are we suppose to do?"

He yelled back, "Charge the fire!"

As we turned towards the firing again, the sergeant was shot in the head, just between his eyes.

I hollered at Wally, our M-79 carrier, to take my rifle and I started pulling the sergeant back to the rest of our company behind the dike. I dove backwards, letting him land on my chest, then dove backwards again and again until I reached the dike.

Once there I laid him lengthwise with the dike and had Wally hold his head out of the water. I went under him with my head and shoulders. Each time I came up the sergeant was rolled up the side of the dike until he was over the top where the rest of the company was.

I rolled over the dike myself and just lie there a minute. Then Wally handed me my rifle and I started shooting again.

Before long I was almost out of ammunition; I had fired all my sixteen clips. I always tried to keep the clips full and ready, plus two hundred rounds I kept in a claymore bag.

As I put my last twenty rounds in a clip, I noticed the first sergeant and a M-60 machine gunner, of another platoon, pinned down on the left flank. They were lying there behind the dike.

I raised up to run over and help them. I yelled at Lupe, to tell him. He grabbed me and pulled me back.

My head had just gone down when a bullet parted the mud where I had just raised up.

Lupe said, "That could have been your head."

I was just about to make another try when a chopper came in, so I helped to load the dead and wounded until almost dark. Looking out front I noticed the first sergeant and the machine gunner were still pinned down.

Wally and Lupe were helping load the wounded and dead. Since the dead were beyond help, we loaded them only if there was room. The rest of the dead we kept on air mattresses floating on water behind the dike, until there was room in the choppers to get them out.

We finally finished. Wally, Lupe and I were just lying there.

I turned to Wally and asked him if I could use his M-79 grenade launcher.

He quickly replied, "Sure."

I get several rounds and stared laying them on the jungle line where all the firing seemed to be coming from.

Looking off to my far left, I noticed the first shirt and the machine gunner were still pinned down.

I turned to give Wally back his M-79, looked at Lupe and said, "I'm going to go get them this time."

He said, "You better stay down or you'll get it for sure." He reminded me about how the sergeant I pulled out had gotten it in the head. I didn't know the guy was dead.

Anyway, so I don't get shot by our own men, I told him to pass it down that I'm going across in front of the company to get them before it's too dark.

Jumping up, I started running as fast as I could in the mud and water. I reached the first shirt and machine gunner and dove in where they were, behind the corner of the dike.

As I was attempting to catch my breath I told the machine gunner to give me a hundred-round belt and I asked him to give me his M-60 and take my rifle.

He handed the machine gun to me and I checked the hundred-round belt in it to make sure there wasn't a short round or a metal clip shoved back. I saw there wasn't any so I attached it to the other hundred-round belt.

"As I start shooting you get up and run back to the rest of the company," I said as calm as I could.

I threw the two hundred rounds over my shoulder and started firing at Charlie's position.

The first shirt and the machine gunner took off as fast as they could run and I was right behind them, firing that machine gun and running backward in that rice paddy slop.

Somehow I managed to make it to the company without falling. Even though I didn't fall I was covered from head to toe with mud from the bottom of the rice paddy.

Since it was getting dark I decided to stay on the left flank of the company. I had run for the left flank because it was straight behind the first shirt and the machine gunner. I moved out to the very end to where I was the last man on the left.

Rice paddies were full of blood-sucking leeches. Lying there in the water I knew I couldn't get the leeches off after dark; if I lit a smoke to burn them off Charlie would spot it. He was just a few yards away.

I knew that because the two men just next to me asked if some more of our men were out there. They heard somebody saying, "Over here, Joe." I told the two guys it was Charlie lying out there trying to get us to leave our positions.

I asked them if they were going to sleep. With obvious surprise they said, "You crazy, man?"

I told them, "That's good. I'll have somebody to guard me while I'm sleeping."

One soldier said, "You must be crazy!"

"No," I said, "I can die in my sleep just as easy as I can die awake. But just remember that's Charlie out there, not our own men saying, "Hey, Joe, over here." With that warning I figured I had two fully alert guards.

Sleeping in all that water wouldn't work so I told them to pass it down to the men on their right that I was going out front for some bundles of rice stalks to sleep on.

I jumped and ran out front and grabbed four or five bundles of stalks to put in the water behind the dike where we were.

When I dove back over the dike they looked at me and once again said, "You are nuts."

I said, "Maybe; wake me up if Charlie comes crawling up. I had one more twenty-round clip, and a bayonet on my rifle.

I laid back, put my steel pot on the dike and shoved the bundles under my body to hold me up just above the water. As I lie there I was thinking over the things that had happened that day.

First, my friend Ray getting hit five times with rifle fire and being temporarily blinded by a bullet that hit his right eye. Luckily he got on the last helicopter taking out the wounded. The brass refused to bring any more in because firing was too heavy at the time.

Then I thought how great those phantom jets looked when they were scraping Charlies' bunkers along the jungle line.

Then I started thinking of when one of the Magnificent Six got killed. He got shot in the head. It didn't seem real that Gary was dead; he had been one hell of a soldier.

Thoughts flashed of his people. I thought of the other dead and wounded I had helped load during the heat of the fire fight.

Also I remembered how my rifle temporarily jammed with mud and how I had just run my cleaning rod down the barrel, just enough to see daylight through it. And how I turned and started shooting again.

The last thoughts were of dragging the dead and wounded to the choppers and loading them, and finally, running out to get the first shirt and machine gunner.

With all that, I was fast asleep. I was awakened the next morning with the sun shining in my face.

We got orders to start moving into the jungle. We found a charred base camp of the Viet Cong. We checked it out and were ordered to move on a company sweep to a place where we could set up a temporary base camp in the jungle.

As we set up camp dark closed in. We had a short break then our platoon lieutenant came by and said we would be going out on a night ambush.

The next day we came back to base camp and were told we would be going back to our main base camp, Cu-Chi, for replacement supplies.

14

One Battle Too Many

After our return we were told we would be heading out toward the Black Virgin on a company sweep. We got our supplies and joined a convoy.

In a few hours we were unloaded alongside the road and waiting for orders. Our lieutenant told us we would be lead platoon. That meant I would be point man again. As soon as the lieutenant finished reading the orders, I moved out.

It seemed like the day was never going to come to an end. Along about late afternoon I started across an opening and had just come into some jungle when I stopped cold.

I stood there, waiting for Lupe to move up to me. As soon as he saw me he stopped.

I told him to go back and stop the rest of the platoon and company. "Tell them we are in a mine field."

He was also to warn the lieutenant we better get out of there and go around the other side. I waited there.

When Lupe finally came back, he said the lieutenant asked how I knew about the mines.

Since I figured he wouldn't believe that the chills and goose bumps were telling me, I showed him a grenade in a hedge just in front of me, that I would have set off if I had kept going.

I told Lupe to make sure the men spread out and didn't bunch up. I knew, and Lupe knew, that all those new men would be scared after the last battle and would feel more secure next to their buddies—that is until they saw men hit mines and their buddies alongside them blown up.

Lupe left again and I waited for him to come back and let me know what battalion brass told our company commander and then our lieutenant.

He came back and said, as usual, we would have to search out the mines. Those were orders from battalion.

Lupe said three men plus him and the lieutenant would be up to help me.

Carefully I moved up through the jungle, checking for more booby traps, and for Charlies. In just a few steps I came out of the jungle onto a well-traveled trail.

I came to a quick halt and stood there waiting for the other men. Within a few short minutes Lupe, Gillman, Lieutenant Gravel and a sergeant that had taken Pat's place, all showed up.

The lieutenant told us to put our bayonets on our rifles and probe the ground with them where we think a mine might be.

I looked at the lieutenant and said, "You guys do that just because those are orders from battalion. That's bullshit. Those guys at battalion ought to come out here and probe with *their* bayonets, until one of *them* is blown sky high."

After that the lieutenant said, "Let's check out the other side of the hedge on the trail." We found a whole Viet Cong village, but it was deserted.

We checked further and found a tunnel. I removed

my web gear and took a forty-four machine gun into the entrance.

I found all kinds of tunnels running off to the side of the one I was in. I was almost too big; it was all I could do to crawl forward.

After a few minutes I wished I had refused to go in, but I had to finish crawling through until I found a way out.

Then it occurred to me if I did run into Charlie that damn forty-four would blow my eardrums out.

Crawling on I kept telling myself, "Keep your cool, don't panic."

After several minutes, that seemed like an hour, I found a trapdoor where I could get out. I slowly opened it, peeking out. I thought, "What if one of our own men saw me and started shooting."

Being real careful, I opened the trapdoor and crawled out. I found I was past the huts and back across the trail. I was in a jungle thicket.

I went back to the entrance of the tunnel and waited for the lieutenant and the sergeant. I gave them a report of the tunnels and where I came out. I told the lieutenant I figured the side tunnels went to the huts because that was the direction they went from the main tunnel, and about the candles along the tunnel walls.

He called for one of the platoon men who carried C-4 for blowing up things. I took the C-4, some caps and a detonator, removed my web gear, crawled a little way into the tunnel and blew up the entrance.

Then we started checking out the huts.

Charlie knew what he was doing. He had booby traps all over. Our lieutenant was the first in our platoon to get hit. About the same time several men in another platoon were wounded.

We put the lieutenant on a stretcher. Several of us packed him around to the trail and out to an opening. The platoon's radio man called a chopper in to get him and the other men that had been hit by booby traps.

The R.T.O. had called for a chopper from Saigon but a medivac chopper that was closer said he would come in and pick up the men. The R.T.O. called the medivac from Saigon and told him he could turn back.

But before he got all the words out of his mouth, a couple more GIs hit a booby trap and the Saigon helicopter pilot was told we needed him.

No sooner was the message sent than the closer copter landed. We quickly loaded the lieutenant and the other wounded men.

I turned to the left towards the open field and headed back toward the trail going around the hedge to Charlie's grass huts.

As I got to the trail I ran into a sergeant who had kept my rifle while we packed the wounded lieutenant. The sergeant was headed for the opening, so I turned back toward where the chopper landed and asked the sergeant where he had put my rifle. He said he had Private Ross take care of it until I returned.

About that time, a mine blew about six feet in front of me. The force of the explosion blew me back ten to twelve feet, into the tall jungle hedge on the other side of the trail.

I briefly saw myself going out in a white light, whiter than anything I had ever seen. I knew I must have died.

Then, all of a sudden, the light went away and I knew I was back in this world.

Realizing this, I got up out of the hedge to see if I still had my arms and legs. Then I fell forward and started crawling on my legs and hands. My eyes hurt too much to open them.

Somehow, without opening my eyes, I had walked back to where I had been standing. Then I realized I had better stop before I blundered into another mine or booby trap.

To be sure where I was I opened my eyes just enough to see. I didn't try to open them very wide because they were full of slivers and hurt so.

After seeing for sure that I was back where I had been, and seeing three men that were blown apart, I closed my eyes to relieve the pain.

A few seconds later Marty, who had been close by, came up and said, "You better lay down, Bud; you have blood all over your chest."

I managed to open my eyes again. I looked down and saw blood on my chest where my fatigues were torn apart from the sharp metal and whatever else Charlie could put around the mine.

I stood there a bit, then I said to Marty, "I could make it to the chopper when it comes in if you will lead me." It still hurt to open my eyes.

Marty agreed to help me.

I laid down to rest. Before long I heard the chopper coming. Marty helped me up and led me to the chopper.

The medivac men jumped out and loaded the chopper full and quickly shut the doors. As I lie there I remembered how different the chopper sounded when the doors were closed. I'd never ridden any choppers that I didn't plan on jumping out as soon as it touched down.

It seemed like only a few minutes until they unloaded us at the hospital in Cu-chi, our base camp.

15

Hospital, Medals and Out

It occurred to me that people were running everywhere. I really didn't know, but it sounded like it. Then I heard someone saying, "He's going into shock!" And that was when they cut my jungle fatigues and boots off of me.

The next thing I remember was three days later. I woke up in bed with tubes running down my nose, stuck in my arms and it seemed like everywhere. That afternoon two or three doctors came by and checked me out. They were the surgeons who had operated on me.

They told me they had taken bits of shrapnel out of me. And the doctor said the reason it hurt to open my eyes was because I had slivers from the mine in them, just like getting slivers in your hand. He said they would be taken out as soon as I was able to get up; I had to be able to walk to his office.

I told him and the head surgeon, "I will be over tomorrow or the next day."

The head doctor said, "You won't be up for at least one week, if not for two weeks."

"I will be getting up tomorrow," I promised. "so you better be ready to pull those tubes out of me or I'm going to do it myself."

They left, just shaking their heads.

The next morning they all came back and the head doctor told the nurses to pull the tubes out. He left for awhile and came back later that afternoon and removed the tape from my stomach.

He was a little mad at me because I proved him wrong about when I would be ready to get up, so he took the tape off as slowly, and as painfully, as possible.

I didn't say anything until he was done and after the eye doctor was through talking to me. Then I turned to the eye doctor and told him I didn't want that head doctor to ever touch me again.

In fact he better not even come near me again without my permission. I explained to the eye doctor that I considered the method used to remove the tape was uncalled for, and someone better tell the head doctor what I said.

I would have told him myself but he left while the eye doctor was talking to me. Anyway, the eye doctor said he would deliver the message. He apologized for the head doctor and said that they see so much in these hospitals that they sometimes forget they are working on people. He then left to finish his rounds.

The next day I went over to his office to get the slivers pulled out. He was very gentle and only did a couple at a time, so I had to go back later. In fact, it took two more visits to get all the slivers out.

He said I was the luckiest man he had ever seen, because if the slivers had gotten into the pupil I would have lost my eyesight.

I felt so much better; now I could actually open my eyes without that awful pain. Also, my stomach wasn't hurting so much when I walked.

About two weeks from my arrival, I was being transported to Saigon by chopper to catch a C-141 to the 249th General Hospital, near Tokyo in Japan. I spent several weeks there, resting and getting my strength back.

On the third or fourth week, the doctor started to cut my orders to go back to Vietnam, which is where I wanted to go.

My ears had been buzzing all the time; it was so bad, sometimes I couldn't hear. I didn't want to tell the doctors about it but finally I figured I better. The doctor checked my ears and said, "You won't be going back to Vietnam because your ear drums need to be repaired."

When he started cutting orders for me to go back to the States, the doctor asked where I would like to go. I told him I'd like to go back to Fort Ord, California. The next week I was loading on another C-141 and headed for Fort Ord.

In California we landed at Travis Air Force Base. As I got off the plane memories flashed back to how my friend and me had just about missed the plane to Vietnam when we jumped the fence for a couple of days.

Then I wondered how many of that bunch on that plane were still alive.

As I walked across the airfield to a bus waiting to take us to Fort Ord, I said a little prayer for the men still over there in Vietnam.

All I could think of was, "God, be with them and keep them safe. I ask this in your son's name. Amen!"

I was thinking of how that awful war could be over if they really wanted it to be over.

For the next few months, I worked with my doctors in the clinic. They were great doctors to work with; they gave me all the convalescence leave I wanted. I would just get back from one leave and they would give me ten, twenty, or just about any number of day's leave I asked for.

Working with the doctors was interesting. I would give audio tests for them, and keep the tools they used for minor surgery sterilized.

Finally they finished the last surgery on my ear, and when I got back from one of my convalescence leaves, the hospital company I was in received orders that some other guys and I were to receive medals.

We went to a meeting where a three-star general handed out medals to us. As I received my medals, a Purple Heart and a Bronze Star with a Valor in Combat device, I couldn't help but think that I had just done what I was supposed to do as an American.

I thanked the general as I saluted him. After it was all over, I walked over to Rick and thanked him for encouraging me to go the the Honor Awards Ceremony and for helping me get ready.

I hadn't really intended on even going to the meeting, but Rick pushed me into going. I also told him, "I don't know why they don't just get this damn war over with." Rick and I had talked about going back again, but I doubt they would have allowed me to re-up.

I would be getting an early out in just a few days. Those days waiting seemed long but soon the day arrived, and none too soon.

I was starting to get real nervous; so nervous I went AWOL two weeks on my last convalescence leave. When I came back I had to face the Old Man. He fined me fifteen dollars and said, "The only thing saving you, Brannon, is your good record." He was referring to my medals and now my early out.

I saluted him and respondedwith a simple, "Yes, sir." I couldn't wait to get going home.

My final "Yes, sir," and salute was when he handed me my discharge papers and told me I was out and on my way home.

After my discharge from the Army I went back home to try and get on with my life. I worked at logging camps, traveling from one to another for many years.

Through those years I have made a lot of changes in my life. One thing that seemed to be missing was God. I found Him through Jesus Christ.

I am married and have three kids. To this day I still travel around, working as a timber faller wherever the better job may be.

I believe that putting God above all things is important because He will take care of you and help show you The Way.